Book #1 of 7 in the celebrated series of solid Biblical principles in great tasting chewable tablets for tween small groups from the unique and marginally funny **ALAN ROOT**

DESIGN-AMENTALS!

[dih-ZYNE-uh-MEN-tuhls]

You're not an accident You're not an animal You're the masterpiece of a Master Mind

"You are worthy, our God,
to receive glory and honor and power;
for You have created all things,
and by your will, they exist."
—Revelation 4:11

ALAN ROOT: BIOGRAPHY

Alan Root has been a children's pastor, rock & roll musicianary to older kids, worship leader, teacher, and writer. He's taller than you think. He enjoys reading books like Treasure Island and Nero Wolfe mysteries and almost anything by C.S. Lewis. He likes microwave popcorn, hiking, playing drums, painting with watercolor, musical techno toys, strong coffee, and basketball. He has a BS in Geology and has done postgraduate studies and teaching at the University of Tennessee. For more than 20 years he's been traveling the US ministering the gospel of Jesus Christ with an electric guitar and the grace of God. He's done music for Wrigleys, Kelloggs, the Billy Graham Children's Crusade, and Dr. Seuss, nominated for a Dove award, won 3 Telly awards, written with Michael W. Smith and has tons of CDs and DVDs as well as books especially for older kids available at **www.alanroot.com**. He has a new book for adults called "ASK MY PASTOR, HE KNOWS THAT STUFF" available at Amazon.com.

He drew the pictures for this book all by himself.

DESIGNAMENTALS!

Written especially for tweens by the perspicacious and loquacious Alan Root!

de•sign•a•ment•als

|dih-ZYNE-uh-MEN-tuhls|

noun 1. A word Alan made up to represent the purposeful creation of space, matter & time
2. the truth of how it is that we exist and why we continue to exist: why it's only rational to know that we do not live in an accidental universe
3. A word your friends struggle to pronounce

We'll use the Bible extensively in our study. Why? Here's why: while Jesus is the perfect picture of God himself, and creation declares his glory; the Bible is God's unique gift to us, the written revelation of God and the primary tool by which we measure truth.

"All Scripture is God-breathed and is useful for teaching, rebuking, correcting and training in righteousness, so that the man of God may be thoroughly equipped for every good work."

—2Timothy 3:16

NOTE TO TEACHERS

HOW TO USE THIS BOOK

This book is perfect for small groups; for reading out loud together and talking about what you just read. **SCHEDULE:** One week for the intro plus two weeks for each of the five chapters adds up to eleven weeks. The last week can be a party with a special guest testimony, cake and ice cream. **DAILY READINGS:** Get a book for each student. Have them write their name in the book and make notes in the wide margins. Strongly encourage them to read a couple of pages a day at breakfast, writing their questions to bring to class. Tell them their daily relationship with Jesus is the most important thing in their whole lives. Those who do their homework can be rewarded with bookmarks you make to remind them of things the Lord impresses on your heart concerning this material. **IN THE GROUP:** Each week pick a student to be a "designated designator" who chooses others to read out loud. Let the conversation be directed by the Holy Spirit. It gets crazy, and sometimes you'll wonder how you got so far off the subject, but it's all good. When we talk with each other, real life happens in a relationship context. And this is great stuff to talk about! Some of the features include: **BLACK BOXES:** jumping off points with simple obvious answers. **Q&A:** Have your kids pencil in their own answers. Then check the first appendix for what I thought the answers were. This is perfect for continual review. **PRAYER:** To let the kids "own" these transformational principles. In the **APPENDIX:** you'll find a section called **HOW TO BE BORN AGAIN:** Read it with your kids and talk with them about belonging to Jesus. Next you'll find **KINETIC ACTIVITIES:** full-contact learning for kids with ants in their pants. The last section is called **DIGGING DEEPER:** for those of your kids who are looking for more. Send those kids to **APPENDIX 4:** and ask them to report back to the whole class on what they discover!

TABLE OF CONTENTS

PSSSST! HEY YOU

If you're sure you've surrendered your life to Jesus, skip this page. But if you're NOT all that sure you belong to Jesus, I suggest you read this page very carefully. It's a matter of life and death.

EVER HAD TO LEAVE SOMEONE'S HOUSE BECAUSE YOU WEREN'T PART OF THEIR FAMILY? THE PROBLEM'S NOT THAT THEY DON'T LIKE YOU. BUT YOU WEREN'T BORN INTO THEIR FAMILY AND THERE COMES A TIME WHEN ONLY THE FAMILY STAYS.

How do you "belong" in this world? Easy. You were born. How will you "belong" in the Kingdom of God? You'll have to be born again.

And it has to happen now. Today's the day. Obey the command of Jesus to follow him and be born again. Head for Appendix 2 where you'll find stuff that'll help you to be born again.

I'm serious. Do it.

WEEK 1: INTRODUCTION

WHY READ THIS BOOK?

This book is an adventure of discovery through worlds of ideas you may never have imagined. If you don't know where you came from, you can't know where you're going.

"The LORD, the Maker of the heavens and earth says this: 'Ask me and I will tell you some remarkable secrets…'" (Jeremiah 33:2-3 NLT)

If all you want out of life is to decide which restaurant you're going to visit tonight and what movie you're going to watch, then drop this book at once—it's not for you! Run away!

Don't say I didn't warn you that this book could be hazardous to a boring future life for you.

Once you figure out that Somebody actually created you and has plans for your life … anything can happen!

"Through him all things were made; without him nothing was made that has been made." (John 1:3 NIV)

ATHEISTS SAY THE COSMOS JUST SORTA HAPPENED ACCIDENTALLY, BY RANDOM, THROUGH IMPERSONAL PROCESSES. DOES THE VERSE ABOVE SAY ANYTHING AT ALL DIFFERENT FROM WHAT THEY SAY?

This is one of the most powerful verses in the entire Bible. To be a Christian and to blow past these words is shallow and careless. Speaking of the Word of God, John says that all things were created by the Word, and identifies the Word as Jesus. As far as the world we live in goes, many believe the ones who say we are the random accident of impersonal processes. There are then two options: either the impersonal processes story is a lie, or the Word of God is. The choice must be faced and the choice must be made.

Have you ever had these thoughts before? Did you know that the Bible declares clearly that God made everything? Did you know that there are highly educated adults that want you to believe God did not do what the Bible says He did?

INTRODUCTION

If I make a drawing, who owns the drawing?

If I write a poem, who does the poem belong to?

If I craft a sculpture, is the sculpture the work of my hands?

If I made the world (and believe me, I didn't!) who would own the world?

This book, Designamentals, is about what happens to us when we discover that God is our Creator. Once we understand that, once we let that thought take hold in our thinking and understanding, it will seriously mess with our lives. There are whole systems of our attitudes that must bow to this one giant, mind-expanding idea: God made me. I am the work of his hands.

Is it MY time? MY life? Do I live for MY pleasure? Should I do things MY way?

If God made me out of nothing with his own hands (read Genesis 1:26 and 27) to whom do I belong? To myself, or to Him?

Don't answer without first thinking...

PRAYER: (You can add your own prayer to this one. Let's talk with God...)

Jesus, I know that you are God and that you made me. I'm not in charge. I'm not the boss. I owe my life to you, my Maker and Master. Let this fact change everything about the way I think and behave.

WEEKS 2 - 3: CHAPTER ONE
THE PRINCIPLE OF CURIOSITY

You're human. What does that mean? What is the essence of humanity?

I suggest it's curiosity. Yes, cats are curious, but they don't care about how to make Portland cement or why the moon revolves around the earth or the formulae for bubble gum. But you and I do because we are human and curious.

Your personal curiosity may have been short circuited by our hyper *entertain-me* society. When I was growing up, my favorite toy was dirt. Tomorrow, as I write this book, people will line up to buy a coupon to let them line up again to buy an expensive silly electronic game. Something has changed. It has affected our capacity for being curious about the world around us.

CHAPTER 1: THE PRINCIPLE OF CURIOSITY

CURIOSITY IS WHERE IT ALL BEGINS.

CURIOSITY IS THE JOY OF WONDERING ABOUT THE WORLD AROUND US. YOUR LIFE WILL BE ONE BORING DAY AFTER ANOTHER BORING DAY UNTIL YOU GIVE IN TO YOUR NATURAL TENDENCY TO BE CURIOUS.

LOOK AT THE WALL. SEE THE LIGHT SWITCH? WHAT DOES IT DO? HOW DOES IT DO THAT? WHO INSTALLED IT? WHO INVENTED IT? WHY IS IT NEXT TO THE DOOR? WHERE DOES ITS POWER COME FROM? HOW IS THE POWER STORED AND HOW DOES IT GET TO THIS ROOM?

ARE YOU GETTING THE IDEA?!!

LEAF THROUGH AN ENCYCLOPEDIA OR WALK DOWN THE NONFICTION AISLES OF YOUR LOCAL LIBRARY AND READ THE TITLES.

MAKE YOU CURIOUS?

CHAPTER 1: THE PRINCIPLE OF CURIOSITY

There's no such thing as a great answer without a great question. If I say, "Sixty-six" to you, that answer means nothing until you know that my question is, "How many separate books make up our Bible?"

Answers answer *questions*. Sloppy questions make for useless answers.

DO YOU HAVE A SENSE OF WONDER? WHAT DO YOU WONDER ABOUT? DO YOU LOOK UP AT THE NIGHT STARS OR WATCH A WORM CRAWL OR SEE A BABY SMILE AND WONDER WHY THIS WORLD IS THE WAY THIS WORLD IS? IT'S 'WONDER'-FUL. WONDER FULL!

Asking questions is the beginning of a life full of wonder. If you decide to let someone else think for you, they most certainly will. But if you decide you're going to look at the cosmos through wondering eyes, you will be rewarded. God doesn't mind when we begin to be curious about this amazing world he made.

"Call to me and I will answer and tell you great and marvelous things which you do not know." (Jeremiah 33:3)

Who am I? **What** am I? And by far the most fascinating, **WHY** am I?

What's real? What's eternal? What's going on with the world around me? Who's running the planet? Is there anyone at the steering wheel, or is everything chance and happenstance?

Do I matter? Does what I think matter? Does what I do or don't do matter? Am I a speck of dust in an infinite universe that has absolutely no significance?

How am I different from an animal? What is love and where does it come from? Who wrote my DNA? Who decided I would exist?

What is merely an accident and what is purposefully guided in my life?

These are really good questions that need really good answers. Do you wonder about these?

CHAPTER 1: THE PRINCIPLE OF CURIOSITY

Modern science was developed by Christians. Roger Bacon, a Christian, invented the "scientific method" later perfected by Francis Bacon, a Christian. Their theory is observe, theorize, test, repeat. A theory of wonder and curiosity!

All kinds of **superstars of science** had strong faith in God: Faraday, Joule, Kepler, Cuvier, Carver, Edison, Einstein, Pascal, Von Braun, Dalton, Morse, Newton, Kelvin, Pasteur, Mendel ... all believers in God and God's creation, all looked for God's fingerprints, all outstanding scientists who had a passionate curiosity.

WHAT HAVE PEOPLE OF FAITH DONE FOR SCIENCE? DOES SCIENCE HONOR OR RIDICULE PEOPLE OF FAITH THESE DAYS? WONDER WHY?

God loves science. If anyone tries to tell you that science and faith don't mix, tell them people with faith in God populate the history of science. Many of science's great champions and heroes were people of faith in God. Say it nicely, but firmly.

If they're as open minded as they claim, they'll be open to the possibility that faith in a God who reveals secrets is crucial to the work of science!

God is not afraid of our questions, he loves curiosity. How do you suppose God feels about answering our questions? He wants us to come to Him with our curiosity. He can't wait to reveal himself to you and me.

"Ask and it will be given to you; seek and you will find; knock and the door will be opened to you. For everyone who asks receives; he who seeks finds; and to him who knocks, the door will be opened." (Matthew 7:7-8)

"Call to me and I will answer you and tell you great and unsearchable things you do not know." Jeremiah 33:3)

Psalm 8 says, **"When I look at the heavens and all that you created, what is man that you should even think of him?"** If Almighty God is thinking of us, shouldn't we be thinking of him too?

The Bible warns us not to mess with our Creator. Don't believe me? **"Destruction is certain"** says the verse on the next page...

"Destruction is certain for those who argue with their Creator. Does a clay pot ever argue with its maker? How terrible it would be if a newborn baby said to its father and mother, 'Why was I born? Why did you make me this way?'" This is what the LORD, the Creator and Holy One of Israel, says: 'Do you question what I do? Do you give me orders about the work of my hands? I am the one who made the earth and created people to live on it. With my hands I stretched out the heavens. All the millions of stars are at my command.'"

Isaiah 45:9-12 NLT

Some call the Genesis creation story a fairytale. How do they know that? Were they there? Are they smarter than God? If they deny that God made the world, will they be in trouble when they see God face to face? Will God maybe ask them why they "argued with their Creator" like the verse we just read talked about? Should they show God respect?

"But God shows his anger from heaven against all who push the truth away from themselves. From the time the world was created, people have seen the earth and sky and all that God made. They can clearly see his invisible qualities —his eternal power and divine nature. So they have no excuse whatsoever..." (Romans 1:18-20 NIV)

There are people who call the Bible's account of Creation crazy. The craziness is in their heads! It makes all the level-headed sense in the world to observe that the obviously designed world around us has a Designer. God has revealed the truth in the Bible and in the world around us. It's so obvious the verse above says the only way to miss it is to "push the truth away."

And it's not wise to argue with God.

CHAPTER 1: THE PRINCIPLE OF CURIOSITY

SINCE SCIENCE IS NOW FULL OF UNBELIEVERS, SHOULD CHRISTIANS WANT TO BECOME SCIENTISTS?

ARE YOU KIDDING? ABSOLUTELY YES!

CHRISTIANS SHOULD BE IN TOUCH WITH THEIR WORLD IN EVERY WAY: ART, CULTURE, POLITICS, PHILOSOPHY, AND SCIENCE TO NAME A FEW WAYS. OF COURSE, IF YOU LOVE BOTANY AND BIOLOGY AND PHYSICS AND ASTRONOMY, PLUNGE IN.

JUST KNOW THAT AS A CHRISTIAN IN TODAY'S SCIENTIFIC COMMUNITY YOU'LL BE AN OUTCAST. WHEN PEOPLE CALL YOU NUTS DON'T BE SURPRISED. EXPECT YOUR FAITH TO MAKE TROUBLE FOR YOU IN THIS DARK, GODLESS WORLD. JESUS WARNED US, "IF THE WORLD HATED ME, IT WILL HATE YOU TOO."

WHAT AN HONOR TO BE HATED BY THE SAME PERVERSE, MIXED UP WORLD THAT HATES JESUS!

We have now begun to explore the joy of curiosity. We've said that curiosity is the essence of being human. We've said that it is the starting place for a life that can't wait for each new day. We've said that science owes much of its advancement to those who wanted to find God's fingerprints in creation.

Modern science ought to be open and curious about God's involvement with His world.

It's not.

SO THE ONLY QUESTION IS: WHICH PART OF GOD'S CREATION WILL YOU CHOOSE TO INVESTIGATE? HOW ABOUT IT? WHAT FASCINATES YOU AS YOU LOOK AROUND? WHAT IN THE WORLD INTRIGUES YOU AND CALLS TO YOU TO LOOK CLOSER? WHAT MYSTERY IS YOUR MISSION?

If you can't be curious about something, don't you find THAT to be curious?

I want to always wonder and be amazed as I seek to find out more about God in the world that he created. I know you do too.

Q&A 1

Q. Why are we often bored? What is the source of our boredom?

____
____
____

Q. What is the antidote for this boredom?

____
____
____

Q. Why should we be curious?

____
____
____

PRAYER (You can add your own prayer to this one. Let's talk with God...)

Jesus, would you awaken in me a curiosity about your creation? What part shall I investigate? I want to ask, seek and knock. Thanks for making such a wonderful, wonder-filled world.

WEEKS 4 - 5: CHAPTER TWO
THE PRINCIPLE OF IMPACT

The word impact can be translated "how it affects your life." If, by the end of this chapter, I have not convinced you that this stuff will impact your life, I'll eat an old pair of gym socks. A figurative pair. Pay attention!

Around 30 AD Jesus taught his disciples a basic formula called, **"Love your neighbor as yourself."** Simple idea.

Around 1850 AD Charles Darwin wrote a book called THE ORIGIN OF SPECIES, with the formula he called "The survival of the fittest." Also not hard to understand. From the title of his book, he said life came from the simple survival of fit organisms over unfit ones. **Don't love your neighbor, survive him.**

So which is it? Is it "Love?" Or is it "Survive"?

Since they oppose each other, they can both be wrong, but **they cannot both be right**. Someone is correct and someone is incorrect. My money is on Jesus over Darwin!

CHAPTER 2: THE PRINCIPLE OF IMPACT

WHY SHOULD WE CARE ABOUT WHETHER GOD MADE US?

SO WHAT IF WE'RE NOTHING MORE THAN A HAPPY ACCIDENT OF A UNIVERSE WHICH ITSELF IS A HAPPY ACCIDENT?

OUR FAITH DOESN'T DEPEND ON WHETHER WE THINK GOD CREATED US DOES IT?

I WANT YOU TO THINK ABOUT THIS. IF GOD DIDN'T MAKE THE UNIVERSE, HE'S NOT WHO HE SAYS HE IS.

IF GOD DIDN'T MAKE THE UNIVERSE, THE BIBLE BEGINS WITH A LIE.

IF HE DIDN'T MAKE THE UNIVERSE, WHAT ELSE HAS GOD CLAIMED THAT'S JUST NOT SO?

THIS IS SERIOUS STUFF. DARWIN MADE UP A STORY THAT IS IN DIRECT CONFLICT WITH THE BIBLICAL ACCOUNT.

WHO IS RIGHT? WHAT IS TRUE?

The truth is, if you believe in Darwin, you'll struggle to believe in Jesus. Did you know that?

Moses wrote the Biblical account of creation. Jesus testified that Moses got it right when he said, **"If you don't believe what Moses, wrote, how can you believe me?" (John 5:47)**

We don't need the science of genetics to tell us that offspring within a species can look different. Is it earthshaking news that you look different from your dad and mom? God created creatures with fabulous riches of genetic options. In each creature, he preinstalled all necessary information to produce huge variations within each species.

Oddly, Darwinists say that sun rays are responsible for this enormity of information. That's a weird way to think. Why claim that sun rays (of all things!) did what only God can do? Darwinists are willing to look awfully silly to stand by their claims. Who knew sun rays had such powers?

If what they're saying is true, could sunscreen slow down evolution?

Am I taking this too far? I've heard Christians say, "Let's just love people and not argue about Creation – it doesn't matter." Really.

If someone told you your parents bought you on sale at Walmart, would that matter or not?

God thinks it's a big deal: Creation is the first thing in the Bible.

DID THE WISE MAN BUILD HIS HOUSE ON A ROCK OR ON THE SAND? IS THE FOUNDATION OF A BUILDING IMPORTANT OR NOT? WHY IS A SOLID FOUNDATION SO ALL-IMPORTANT?

The storms of your life will shake the house of your life. If your life is built on the foundational truth that God created you and loves you, your house will still be standing when the storm is over. If you think life is an accident of sun rays, is that a firm foundation? When your life gets hard, will you pray to sun rays?

CHAPTER 2: THE PRINCIPLE OF IMPACT

I may not have convinced you yet that what we believe about creation radically affects our life. Plus there's the matter of me eating gym socks from the first of the chapter. All right, time for some history.

In 1850, Darwin introduced his theory that God did not create the world and all that's in it. From Darwin's idea, Karl Marx wrote the communist manifesto, leading to the wholesale slaughter of millions of devalued human beings. From Darwin's idea, Adolph Hitler believed that some races were not fit to live anymore, and again millions died. From Darwin, Sigmund Freud devised his perversion of psychology, destroying the basic unit of a healthy society, the family. As a follow-up to all this, Nietzsche then declared God to be dead. Darwin, Marx, Freud and Nietzsche poisoned our world with godless ideas

Denying the Creation has crippled the modern world. Not a single scientific advancement has come from Darwin's theory. But a lot of money has been made from it, which is why Darwinism hasn't been laughed into extinction long ago.

All rational thinking begins with this truth: God is our Creator.

We have two choices.

Choice #1: Our daily lives will be simpler if we just knuckle under and go with the Darwin/Naturalist thing. It requires no thinking on our part. We simply buy the crazy stories of exploding gas and protein soup and sun rays and giraffes growing longer necks by space rays zapping DNA. We smile and nod. We just conform. We let the thought police know we pose no threat to the status quo. We agree with the "smart people" because "everybody knows" the Bible is out of date.

Our other choice is much harder.

Choice #2: Seek truth. Think for ourselves. Take our God-given curiosity and put the universe to the test. Does it bear the marks of a Creator? Does it reveal purpose and design? If so, then decide to laugh at the Darwinists, to stand up for intellectual honesty, to believe God. To stay faithful to what we know.

Prepare to be made fun of. To be labeled anti-science, even though science conclusively points to a Creator. To be sneered at by those whose faith is shipwrecked by silly evolution stories.

Then the leaders of the Levites called out to the people: "Stand up and praise the LORD your God, for he lives from everlasting to everlasting!" Then they continued, "Praise his glorious name! It is far greater than we can think or say. You alone are the LORD. You made the skies and the heavens and all the stars. You made the earth and the seas and everything in them. You preserve and give life to everything, and all the angels of heaven worship you."

Nehemiah 9:5&6 NLT

To borrow a phrase from President Reagan, our friends the Darwinists "are not ignorant, they just know a lot of things that aren't so." Can you really believe that life on earth is a cosmic accident of sun rays hitting wet rocks? Sure. You can also believe that 2 plus 2 is 5, but it will catch up to you at some point because it's just not true. Truth is kind of important, isn't it?

WHAT DIFFERENCE DOES IT MAKE WHETHER OR NOT GOD MADE YOU? DOES IT AFFECT THE BIBLE AND ITS CLAIM TO CONTAIN THE TRUTH? DOES IT AFFECT MORALITY AND THE IDEA OF ABSOLUTE RIGHT AND WRONG?

It is better to know the truth. It is better to reject what is not true. It is better to tell others the truth and warn others about what is not true. Truth is light, freedom and power. Truth is health and hope, life and breath.

And one more thing about truth: Truth is Jesus Christ himself revealed in the Bible and creation and the empty tomb. Read John 14:6 where Jesus says it.

Do you know Truth himself? He loves you.

WHAT ABOUT IT? ARE YOU THE MASTERPIECE OF A MASTER MIND, OR JUST THE ACCIDENT OF SUN RAYS SHINING ON WET ROCKS?

I WANT MY LIFE TO HAVE MEANING. BUT HOW CAN IT HAVE MEANING IF THE DARWINISTS ARE RIGHT? THAT MY ORIGIN IS MEANINGLESS? IF GOD IS NOT MY CREATOR, HOW CAN I EXPECT MY LIFE TO MATTER IN ANY WAY?

PERHAPS THAT'S WHY SO MANY PEOPLE IN OUR SOCIETY ARE AIMLESS AND BORED AND LACKING IN PURPOSE. THEY HAVE COME TO BELIEVE THE LIE OF EVOLUTION. THEY THINK LIFE IS AN ACCIDENT, AND THERE'S NO ULTIMATE MEANING TO ANYTHING.

THEY THINK THAT OTHER PEOPLE'S LIVES DON'T MATTER, BECAUSE THEY THINK THEIR OWN LIVES DON'T MATTER.

THEY COULD NOT BE MORE WRONG.

Who cares about Creation?

I do. I want you to know that one reason God loves you is because he made you.

Who cares about Creation?

God does. All of his authority comes from his title of Creator. If God is not our Creator, he is not our God at all. If we are the accident of rocks in hot water, there is nothing that matters.

DOES IT MATTER WHETHER YOU BELIEVE YOU ARE CREATED BY GOD OR JUST AN ODD STRING OF ACCIDENTAL EVENTS?

Who cares about Creation?

I guess now you do. Because you are the one that God decided to create. God really does love you.

Never doubt that for a second.

Q&A 2

Q. When we say that ideas have consequences, what does that mean?

Q. Should we even be talking about Creation? Isn't it divisive? Might we offend someone?

Q. Darwin's theory is just a harmless idea, right? It doesn't affect our lives, does it?

PRAYER (You can add your own prayer to this one. Let's talk with God...)

Jesus, I know that you made me. I know I'll get laughed at because I believe you made the world. I don't care: It's true. I hope my friends can know you made them and that you love them just as much as you love me.

WEEKS 6 - 7: CHAPTER THREE

THE PRINCIPLE OF AUTHORITY

It's fascinating that one person's facts are not necessarily the next person's facts. In politics, sports, weather, theology, to name a few areas of knowledge, different people have different things they call 'fact.' On what authority do we believe what we believe? What makes a fact factual?

HAVE YOU EVER FOUND SOMETHING WAS JUST NOT TRUE? HAVE PEOPLE TOLD YOU THINGS YOU KNOW ARE JUST NOT SO? HAVE YOU EVER TRIED TO PRETEND YOU KNEW MORE THAT YOU ACTUALLY DID?

People make things up. They try to make you think they know everything. Our immediate response when we hear anybody claim anything whatsoever should be, "How do you know that?" How do we recognize truth when we come across it? How can we say for sure something is reliable?

In other words, how can we possibly know what to believe or disbelieve?

CHAPTER 3: THE PRINCIPLE OF AUTHORITY

Let's get something straight. There's reality and there's theory. They are not the same.

For example: I don't care what climate scientists claim, they don't KNOW what the weather will be in 25 years. I mean, they can't tell you what the weather will be this afternoon! They're guessing (or lying), but they can't possibly know. Their best guess may be right and it may be wrong. They have theories, not reality.

When we use theories, we're saying, "This is our best guess based on what we think we know right now." In stark contrast, reality means "what actually is, regardless of anything we think we know or don't know."

When a Darwinist looks at fossils, they make up stories, then call those stories "facts." But they are not reality, they are stories, in other words, theory. But Darwinist claims that if you don't believe their "facts," you're anti-science. Huh? The rest of us look at those fossils and see dead animals that became rocks. Is a made up story reality, or theory? Is a theory the same as fact?

Do Darwinists know the difference between truth and theory? I don't think they do.

CHAPTER 3: THE PRINCIPLE OF AUTHORITY

DID YOU KNOW THAT CHARLES DARWIN ALWAYS, ALWAYS CALLED HIS THEORY "A THEORY?" I'VE HEARD MANY PEOPLE SLOPPILY CALL DARWIN'S THEORY A "LAW." THEY CAN'T TELL WHAT'S ONLY THEORY AND WHAT IS DEMONSTRABLE SCIENCE.

NO ONE HAS EVER SEEN A SPECIES MUTATE INTO A NEW SPECIES. EVER. THERE'S NO PROOF OF EVOLUTION IN THE FOSSIL RECORD EITHER, OR EVOLUTIONISTS WOULD JUST REPRODUCE THE PHOTOS INSTEAD OF DRAWING ALL THEIR DRAWINGS. SO IF NO ONE HAS EVER SEEN IT, WHY DO SOME PEOPLE SAY IT'S A FACT?

NO MATTER HOW LOUDLY AND REPEATEDLY DARWINISTS CLAIM THEIR THEORY IS AN ESTABLISHED FACT, THE REALITY IS THAT IT'S JUST NOT.

SO, A VERY INTRIGUING QUESTION ARISES: ARE EVOLUTIONISTS "ANTI-SCIENCE?" ARE THEY SUBSTITUTING THEORY FOR REALITY?

Here's another important distinction in wording. Listen up. The word 'evolution' has two very different meanings. Two. Different. Tricky, huh?

CAN A CAT HAVE PUPPIES? CAN A COW HAVE COLTS? CAN A HEN HAVE DUCKLINGS? THE BIBLE IN GENESIS SAYS GOD CREATED LIFE IN 'KINDS.' ANOTHER WORD FOR KIND IS 'SPECIES.'

The word 'evolution' can mean variation WITHIN a species (you're not an exact copy of your parents). But 'evolution' can also mean the beginning of a brand NEW species (your mom is descended from an amphibian in the past). The same word, two radically different meanings. One use of the word evolution is verifiable science. The other use of the word evolution is silly nonsense! But do people realize this?

When people say, 'evolution,' assume they mean Darwin's theory, set out in his book "Origin Of Species." Darwin's theory claims new species have 'evolved' from other species.

It's intentionally misleading. Evolutionists want you to think new species evolve all the time because you don't look exactly like your parents!

Changes within a species constantly take place due to mixing of the parent's genetic information in reproduction. Happens all the time and it's demonstrable science.

But it's incredibly sloppy thinking to take that simple scientific observation and then deceitfully twist it to "prove" a theory that life came from rocks in hot water. To believe that life is an "accident" is crazier than believing we can create a desktop computer by dropping a bunch of laptops! Don't try it, it won't work and you'll be in a lot of trouble!

God constructed living things by processes we cannot begin to understand. Those creatures live and replicate in ways we can barely guess. Yet Darwinists tell us life is nothing more than undirected random recombinations of mutated genes. It's comical. Call it "frog to dog" evolution. Call it "space ray" evolution. Call it "my great grandfather was protein soup" evolution. Small changes within species are observable science. But that doesn't prove that life arose from rocks in hot water as the Darwinists claim. If the smartest person in the world says life began as rocks in hot water, I give you permission to love them and laugh politely!

Christianity is made up of individual Christians. In the same way, science is made of individual scientists. Just as Christians can give Christianity a bad name, scientists can misrepresent science and give it a bad name.

Science is the systematic study of the natural world by using observation and experimentation. To get noticed, some scientists will sometimes make crazy unscientific claims. Then others looking for whatever's new will go along with that, and what oughta be laughed at winds up in textbooks.

WHAT MAKES SOMEONE AN EXPERT? IS AN EXPERT EVER WRONG? EVER MAKE MISTAKES? HOW MUCH TRUST SHOULD WE GIVE THEM?

Does God disagree with **science**? Of course not: God made the laws of science.

Does God ever disagree with **scientists**? Oh mercy, yes!

Who gets to be right then?

"Christ is the mighty power of God and the wonderful wisdom of God. This "foolish" plan of God is far wiser than the wisest of human plans... God deliberately chose things the world considers foolish in order to shame those who think they are wise... God chose things despised by the world, things counted as nothing at all, and used them to bring to nothing what the world considers important, so no one can ever boast in the presence of God."

1Corinthians 1:24-29 NLT

In discovering truth, science is limited.

Say a basketball rolls across the floor. What can we discover from scientific observation? We can measure speed, velocity, sphericity, displacement, mass, surface irregularities, and plot a vector analysis. From this point on, we're guessing. Anyone who tells you more than this is simply telling their "story." Did the ball fall off a shelf, or did a coach whose job is on the line roll the ball onto the floor during practice after a 20 point loss? Is the ball on the rebound from a shot? Was the shot made or missed? Did the shot occur in the first quarter or did it win the game?

Did that basketball 'evolve' from a sleeve of golfballs hit by cosmic rays from the popcorn machine?

Scientific observations only reveal so much.

Where did life come from? Can we use a thermometer, a fossil or a telescope to explain the origin of the universe? I'm saying scientists don't know everything! And we should not look on them as more reliable than we ourselves are!

CHAPTER 3: THE PRINCIPLE OF AUTHORITY

SCIENTISTS ONCE BELIEVED THAT THINGS BURN BECAUSE OF AN UNSEEN SUBSTANCE CALLED PHLOGISTON. I'M NOT JOKING. THEY REALLY DID!

SCIENTISTS BELIEVED THAT IF SOMEONE WAS SICK, THEY HAD TOO MUCH FLUID, SO THEY DRAINED BLOOD FROM PATIENTS. NOT KIDDING. THAT'S WHAT THEY THOUGHT!

SCIENTISTS THOUGHT LIFE ORIGINALLY WAS PRODUCED FROM INORGANIC COMPOUNDS BECAUSE THEY OBSERVED FLIES HATCHING OUT ON GARBAGE. FOR REAL. THERE WAS THE GARBAGE AND THERE CAME THE FLIES, SO THAT'S WHERE LIFE CAME FROM: GARBAGE PILES. THEY WERE SERIOUS!

SCIENTISTS ONCE BELIEVED THERE WAS NO REASON FOR DOCTORS TO WASH THEIR HANDS BETWEEN PATIENTS. EWWWWW.

SUPER SMART SCIENTISTS...CAN BE EVERY BIT AS KNUCKLEHEADED AS YOU AND I.

This is why the principle of authority is so important. This is why we should always ask the question we learned earlier: "How in the WORLD do they know THAT?!!"

WHO SPEAKS WITH AUTHORITY IN YOUR LIFE? IS IT SOMEONE WHO DISAGREES WITH WHAT GOD SAYS IN THE BIBLE? ARE THEY RIGHT? IS GOD WRONG?

"The crowds were amazed at [Jesus'] teaching, for he taught as one who had real authority." (Matthew 7:28-29 NLT)

God is smart and he made smart people, but smart people can't be God, and they can't be smarter than God. When you read your Bible, you get God's wisdom! Way better than smarts!

What's the source of your information? Where does your knowledge come? If your knowledge disagrees with God, what then? God is the final authority on smart and wise. Be warned!

"What an evil, bitter thing it is to forsake the LORD your God, having no fear of him." (Jeremiah 2:19 NLT)

Q&A 3

Q. What is scientific law? What is theory? Does a scientific theory become a scientific law simply by repeating it over and over and drawing nice pictures?

____

____

____

Q. What are the two totally different definitions of the one word "evolution?"

____

____

____

Q. What's the question you should ask of yourself when someone claims something that seems a little loopy like "The universe is nothing more than a lucky freak accident?"

____

____

____

PRAYER (You can add your own prayer to this one. Let's talk with God...)

Jesus, if someone says something that contradicts what you say in the Bible, then I'm going to wonder about them, not you. They could be right or wrong, but you're always right. I trust you most of all.

WEEKS 8 - 9: CHAPTER FOUR
THE PRINCIPLE OF TRUTH

There's a basic principle of logic. If two statements discredit each other, they can both be wrong, but they CAN'T both be right.

This is the principal of truth.

Our ideas about the world around us are, in the end, right or wrong. If evolution says the universe is an impersonal accident and creation says God personally made everything, those ideas can both be wrong, but both cannot be right.

According to logic, that is.

To get around this, some good people foolishly attempt to convince everyone that the Bible and evolutionary theory don't contradict. They either don't know evolution or they don't know the Bible. Because...well, lets look at why they DO contradict in huge ways and see if they are in any way compatible.

CREATION SAYS THE WORLD WAS MADE IN 6 DAYS, EVOLUTION SAYS 8 BILLION YEARS. CREATION SAYS PLANTS WERE MADE BEFORE THE SUN, EVOLUTION SAYS THE SUN CAME FIRST. CREATION SAYS THE HEAVENS WERE MADE ON DAY 4, EVOLUTION SAYS THEY'VE JUST ALWAYS BEEN. CREATION SAYS WHALES WERE MADE BEFORE LAND ANIMALS, EVOLUTION SAYS WHALES CAME FROM LAND ANIMALS. CREATION SAYS GOD STARTED WITH TWO PEOPLE, BUT EVOLUTION SAYS IT WAS MICROSCOPIC PROTEINS TO FISH TO FROGS TO APES TO US. CREATION SAYS WE'RE NOT ANIMALS, EVOLUTION SAYS WE ARE. CREATION SAYS SPECIES ARE GOING EXTINCT, EVOLUTION SAYS THEY'RE POPPING UP EVERYWHERE ALL THE TIME.

I COULD GO ON AND ON: THERE ARE ACTUALLY ALMOST NO SIMILARITIES BETWEEN CREATION AND EVOLUTION. THEY REALLY DO CONFLICT! I WASN'T MAKING IT UP! CAN BOTH OF THEM BE TRUE?

Evolutionists are "Naturalists." They believe there is no God, only the natural, physical world we can see. They believe in the circle of life. They believe in Mother Nature and Mother Earth. They don't allow miracles.

Evolutionists think science explains everything with no mystery. They think the laws of physics are supreme (except for the 1st and 2nd laws of Thermodynamics which disprove everything they believe). They say God is a myth, a crutch, a superstition. They believe in "The Force" but not in Luke! (Alert! Star Wars/Gospel joke!)

THE SECOND LAW OF THERMO-DYNAMICS SAYS EVERYTHING IS WINDING DOWN. BUT EVOLUTIONISTS SAY IT'S GETTING MORE HIGHLY ORGANIZED. WHICH IS TRUE?

Naturalists want a world they can control. They want God in a box so he can't get out and mess with their wrong behavioral choices. They want a world without God where they can do as they like with no consequence and no moral law. A natural world where there is no right and wrong.

But God doesn't play that way.

CHAPTER 4: THE PRINCIPLE OF TRUTH

I'd like you to think about something. Science is the discovery of God's laws like gravity, entropy and electromagnetism. Naturalists believe God's laws can take God's place so that whether or not he exists is irrelevant. I'm not joking, they do.

They believe the universe can exist and run all by itself. God can take a break, go on vacation, grab a nap, or even not exist because the laws of science are watching over the universe. Don't laugh!

If you came upon the US Constitution lying on a table would you believe a hairless animal fell on the table and got mashed by a meteor and flattened by an elephant's foot to make a piece of parchment and then ink was squirted by an octopus making shapes that look like letters and freakishly make sense to us? Just accidentally? No? Are you sure? Evolutionists would.

Scientific laws reveal a Law Maker. If there is no Law Maker, there are no laws. No Creator, no science. It's the only logical possibility.

For a naturalist to be right, nonsense has to make sense. They've ditched their faith in God and chosen to believe the ridiculous. Romans 1 says they **"push the truth away from themselves."**

Guess what? Our government is searching the distant heavens for extraterrestrial intelligence. How will they know when they find it?

Language. Perhaps a series of nonrandom beeps or blips, which would be a signal from them. You know: THEM. Oooooooh!

So, according to NASA, wherever we find language, we conclude that there is intelligence. Remember that.

Because our DNA molecules are a language with syntax, punctuation, verbs and nouns? How did we get this language if impersonal, non-intelligent forces supposedly formed life?

WHO WROTE THE LANGUAGE OF DNA? IS IT JUST LUCKY FREAK MUTATIONS OR DID GOD DESIGN IT LIKE THE BIBLE SAYS?

Is the universe random, or organized? Did you know that organization needs an Organizer? That design reveals a Designer? And creation a Creator? Not a difficult concept, is it?!

DNA is a living blueprint of jaw-dropping detail and intricate design, a language written in molecules, with astounding miniaturization and computational power. DNA is one item among many things Darwinists cannot account for. Hey, did you hear about bees?

BEES LIVE IN COMMUNITY WITH WELL-DEFINED SUBDIVISIONS PERFORMING DIFFERING TASKS FOR THE COMMON GOOD, COMMUNICATING PRECISELY BY MOVEMENTS, REACTING AS ONE MIND TO DANGER. OH YEAH, AND THEY'RE NOT SUPPOSED TO BE ABLE TO FLY. CREATED? OR THE RANDOM RESULT OF ROCKS IN HOT WATER?

Guess what? The Earth is a complete miracle: small enough not to crush life, but large enough to retain an atmosphere. A moon just the size to preserve planetary tilt which allows seasons, with a giant planet (Jupiter) nearby for meteor absorption and protection. Orbiting a single star of perfect size, temperature and distance providing energy without cooking us to death or catching us in its gravity. Miraculous. It's as if a uber-powerful, uber-creative, uber-intelligent being had planned and executed the complexity we see all around us. Hmmm.

"Then the LORD answered Job from the whirlwind: 'Who is this that questions my wisdom with such ignorant words? Brace yourself, because I have some questions for you, and you must answer them. Where were you when I laid the foundations of the earth? Tell me, if you know so much. Do you know how its dimensions were determined and who did the surveying? What supports its foundations, and who laid its cornerstone as the morning stars sang together and all the angels shouted for joy?'"

Job 38:1-7 (NLT)

Physicist Richard Morris observed the organization of the cosmos and dryly quipped, "It's almost as if the universe had been consciously designed."

Evolutionist Fred Hoyle said, "The chances of life evolving without intelligent design are comparable to a tornado ripping through a junkyard and producing a Boeing 747."

"THE DANGER WHEN MEN STOP BELIEVING IN GOD IS NOT THAT THEY'LL BELIEVE IN NOTHING, BUT THAT THEY'LL BELIEVE IN ANYTHING." – G. K. CHESTERTON

Remember the old proverb, "He who stands for nothing will fall for anything." It takes more blind gullibility to fall for the rocks in hot water story that it ever did to believe Moses' account of creation in Genesis.

CHAPTER 4: THE PRINCIPLE OF TRUTH

THINK FOR A MOMENT ABOUT THIS QUESTION: IS THERE SUCH A THING AS "ABSOLUTE TRUTH?" "ABSOLUTE TRUTH" IS TRUTH THAT'S SO TRUE, IT'S TRUE WHETHER YOU BELIEVE IT OR NOT. IS THERE SUCH A THING?

LET'S USE LOGIC. SAY THERE'S NO ABSOLUTE TRUTH. IF THAT'S TRUE, IT'S FALSE. BECAUSE IT'S ABSOLUTELY TRUE THAT THERE IS NO ABSOLUTE TRUTH, SO THE ABSOLUTE TRUTH OF THERE BEING NO ABSOLUTE TRUTH IS REFUTED BY THE FACT THAT IT WOULD BE ABSOLUTELY TRUE. YOU THINK THIS IS HARD TO READ, YOU SHOULD TRY WRITING IT!

BUT LOGICALLY, THERE HAS TO BE ABSOLUTE TRUTH. OTHERWISE, WE COULD NOT SAY THAT IT ABSOLUTELY DIDN'T EXIST. AT LEAST, NOT WITH ABSOLUTE CERTAINTY.

IF WE CAN SAY THAT THE LACK OF ABSOLUTE TRUTH IS ITSELF AN ABSOLUTE TRUTH, THEN WE HAVE PROVED THE EXISTENCE OF ABSOLUTE TRUTH!

And since there must be absolute truth, I wouldn't look to mere mortal man for it. I'd look instead in God's revelation in the Bible.

You see, evolution is a theory of unknown substances in an unknown past and an unknown place under unknown conditions producing unknown organisms through unknown processes in an unknown time span. That's a little short of what I'd like to call absolute truth.

"In the Beginning, God created the heavens and the Earth." Genesis 1. Awfully absolute.

"Through Him all things were created, without Him, nothing was made." John 1. Look at the words 'all' and 'nothing!' They too are absolutes.

"You are worthy Lord God, for you created all things!" Revelation 4. Again, absolute.

If I want truth that is solid, dependable, ultimate, and absolute, I'm going to have to expect it from God alone. Only he is reliably trustworthy.

Darwinism is a system of guesswork. There is not a four year old in the world you could sell Darwin to. That's why you need gullible adults.

Q&A 4

Q. What happens when two ideas contradict each other? What about the Bible and evolutionary theory, are they in harmony?

____

____

____

Q. What is naturalism? Why is it intellectually odd for a scientist to be a naturalist?

____

____

____

Q. Is there absolute truth? Is evolutionary theory absolute? How about creation?

____

____

____

PRAYER (You can add your own prayer to this one. Let's talk with God...)

Jesus, thank you for truth that's true whether I believe it or not. This world is full of wild guesses. You satisfy my mind and heart with your truth. I believe that ultimate, absolute truth is you yourself.

WEEKS 10 - 11: CHAPTER FIVE

THE PRINCIPLE OF MEANING

What is meaning? And think, why does meaning exist?

"In the Beginning was the Word," says John 1:1.

The Greek word for beginning is 'logos' -- a title of Jesus Christ. 'Logos' indicates meaning: revelation, word, communication, logic, reason, intelligence, explanation, purpose. Jesus created the universe with meaning and purpose and rationality because that's who he is.

He thought us up and formed us inside our mother's womb and has watched over us our whole life. For our sake, He was born as a baby, died on the cross and rose from the dead. We have a hope and a future. The One who made us loves us, and because of that amazing love, our lives have meaning.

CHAPTER 5: THE PRINCIPLE OF MEANING

DON'T TRY TO TELL ME YOU DON'T CARE WHETHER YOUR LIFE HAS MEANING OR NOT. I DON'T BELIEVE YOU. MEANING IS THE BASIC CRAVING OF THE HUMAN HEART.

A CRAVING SO DEEP IT HURTS.

WE CAN WEATHER THE STORMS AND THE TRIALS AND THE PERPLEXITIES OF LIFE IF WE BELIEVE THAT THERE'S A REASON BEHIND IT ALL, THAT IT'S NOT RANDOM CHANCE WITH NO ULTIMATE PURPOSE.

THE REASON WE CRAVE MEANING IS BECAUSE IT IS THE CHARACTER OF OUR CREATOR: JESUS, THE WORD OF GOD, THE MESSAGE OF GOD, THE ONE WHO IS THE EXACT REPRESENTATION OF GOD.

WE WERE MADE WITH A NEED FOR MEANINGFUL RELATIONSHIPS AND MEANINGFUL CONVERSATIONS AND MEANINGFUL LIVES.

I MEAN IT.

God did not create the planets just to play marbles. He didn't create the sun because it was too dark to read. He didn't create life on earth because He couldn't think of anything else to do.

God has a plan. He moves with purpose. That's why we call Him God. Do we know what the plan could be? We've been given small glimpses, enough to satisfy our questions for now. We know that this world is only just the beginning. We know that God will destroy it someday (it's fatally messed up and shouldn't continue much longer) and then he'll create a brand new one. One that isn't messed up.

How do we get into that new creation? By letting Jesus create us new again. **"If anyone is in Christ, they are a new creation!"** 2Corinthians 5:17.

Have you entrusted your life to Christ? Then you, my friend, are a new creation. And a new creation is automatically part of the coming creation.

Congratulations!

The last written communication we got from our Creator, He described a new creation in Revelation chapters 21 and 22. Just as you have to be born to get into THIS creation, you have to be born to get into THAT one. "Born from above," is how Jesus explained it to Nicodemus in John 3:3. It only makes sense. The door to this messed up world is birth. The door to the new world that's not messed up is *re*-birth. Being born **again**!

In this creation, we have loneliness, wickedness, pain, sadness, sickness, and death. But this creation is passing away and a new one is coming. Because God gives our lives meaning, the sorrows of this world all have a clear purpose: to prepare us for the new creation.

Call this world what you like: training, practice, boot camp. Whatever you call it, it has a purpose because God is at work. No present trouble is wasted time. God doesn't make mistakes.

So when life is hard, we rejoice because we are being prepared for the next thing. The new creation is literally out of this world. So don't fall for the nihilism (look it up) of evolution. Don't be robbed of your faith and hope. God has much more for us than mere "survival of the fittest."

Can you think? Why? Think of why you can think. Is it because a meaningless universe simply exists and you happen to accidentally have exactly the molecular structure to ...

Oh, why even go down that road?! It's ridiculous!!! It's brainless to believe inorganic compounds that exist for no reason somehow produced a brain. Information doesn't come from sun rays or rocks in hot water. Our minds can only be explained by the existence of our Creator's mind. His great mind gave us our little minds.

DO YOU EVER WONDER WHY YOU WONDER?
QUESTION WHY YOU HAVE QUESTIONS?
EVER ASK WHY YOU ASK WHY?

"The fool says in his heart, 'There is no God.'" Psalm 14:1 says we turn into nincompoops when we disconnect our minds from Him.

God's thoughts are full of lovingkindness towards you and me. His mind is brimming with wonderful blessings for our future.

"I KNOW THE PLANS I HAVE FOR YOU," SAYS THE LORD, "PLANS TO PROSPER YOU, TO GIVE YOU A HOPE AND A FUTURE." SAYS JEREMIAH 29:11. WHAT DID GOD JUST SAY?

Choose. Either rocks and hot water became us by freakish coincidence through unknown processes in a purposeless universe; or God intentionally made us — he thought of us even before we were born — that who and what and where we are is not an accident.

Either my life is pointless and empty, or my life has meaning and purpose. I can believe I'm just a jazzed-up primate, or that God crafted me in his own image.

"He is the God who made the world and everything in it. Since he is Lord of heaven and earth...he has no needs. He himself gives life and breath to everything, and he satisfies every need there is. From one man he created all the nations throughout the whole earth. He decided beforehand which should rise and fall, and he determined their boundaries. His purpose in all of this was that the nations should seek after God and perhaps feel their way toward him and find him—though he is not far from any one of us. For in him we live and move and exist."

Acts 17:24-27 NLT

Logic has a Law of Causality, which simply means nothing just happens. The genius Albert Einstein said, "The scientist is possessed by the sense of universal causation." He's saying since something has obviously happened, then just as obviously, something made it happen.

The cosmos is proof that there's a Someone who makes things happen, that He does things with purpose, that the world is not an accident, that there is a Creator. If God didn't make us, we would never have been able to even wonder if he did!

IF WE FIND A BROKEN WINDOW AND A BASEBALL INSIDE A ROOM, AND THE BASEBALL PLAYERS OUTSIDE ARE ALL RUNNING AWAY, SHOULD WE ASSUME THE WINDOW JUST BUSTED ITSELF?

What about the Law of Causality? Is there any chance the universe was not created? Is there no ultimate cause of the cosmos?

EITHER YOU'RE THE MASTERPIECE OF A MASTER MIND, OR YOU'RE THE HAPPY ACCIDENT OF EXPLODING GAS.

IT'S ACTUALLY THAT SIMPLE.

YOU CAN DRESS UP THE EXPLODING GAS STORY WITH DRAWINGS AND CHARTS AND TALL TALES, BUT DON'T TRY TO TELL ME YOU CAN PROVE THE EXPLODING GAS STORY WITH DRAWINGS AND CHARTS AND TALL TALES. A BUNCH OF SILLY NONSENSE WITH CHARTS AND DRAWINGS IS STILL SILLY NONSENSE.

BUT OF COURSE, FEEL FREE TO FALL FOR IT IF YOU REALLY WANT TO. YOU'RE LESS LIKELY TO BE MADE FUN OF IN SCHOOL IF YOU BELIEVE IN THE SILLY NONSENSE OF EVOLUTION.

AS FOR ME, IT TAKES MUCH LESS FAITH TO BELIEVE GOD MADE US ALL AND THE WORLD WE LIVE IN. I CAN'T IMAGINE ROCKS IN HOT WATER DOING WHAT I SEE ALL AROUND US.

When we marvel at a sunset, do we think it's nothing more than refracted light obeying physical laws? Remembering a loved one, is that just fatty tissue synapses that store electrical signals? Watching a documentary on some incredible animal species, is that creature simply the dumb luck of mutant molecules? Time? How did God think up time? But he did. Space, energy, mass, atoms, velocity, cohesion, electromagnetism, galaxies, mitosis, migration, reproduction, symbiosis, eyesight, beauty, chlorophyl, brain waves, synapses, heartbeats, wonder, morality, love, loyalty, and enjoyment are just a few things I DEFY any evolutionist to try to explain using rocks in hot water.

Meaning? How are we able to talk to each other and understand words? There's no such thing as random meaning. Intelligence only comes from intelligence.

We must never for a moment doubt that God made us and loves us more than we'll ever know. The evidence is everywhere we look in his creation, because He wanted it that way. He left us clues of his love for us.

Q&A 5

Q. What does the Greek word LOGOS mean?

Q. What is nihilism? Did you look it up? How is it related to evolutionary theory?

Q. What does the Creator of the universe want from us? How does that work?

PRAYER (You can add your own prayer to this one. Let's talk with God...)

Jesus, I thank you for creating the universe so you and I could have a relationship. You must really love me. I love you back. Because of you, my life has purpose and a future and great meaning.

APPENDIX 1: Q&A

CHAPTER 1: THE PRINCIPLE OF CURIOSITY

Q. Why are we often bored? What is the source of our boredom?

WE MISTAKENLY THINK THE WORLD MUST ENTERTAIN US, AS IF WE WERE KINGS AND THE WORLD WERE OUR COURT JESTER. WHEN THE WORLD NO LONGER ENTERTAINS US, WE BECOME BORED.

Q. What is the antidote for this boredom?

CURIOSITY IS THE GOD-GIVEN HUMAN TRAIT THAT MAKES US WANT TO INVESTIGATE AND ASK QUESTIONS AND FIND ANSWERS.

Q. Why should we be curious?

WE COULD LET EVERYONE ELSE THINK FOR US AND LEARN FOR US AND EXPLORE FOR US WHILE WE WATCH, BUT THAT'S NO FUN. WE LOVE TO LOOK AT THE WORLD THROUGH CURIOUS EYES BECAUSE THAT'S HOW GOD MADE US.

CHAPTER 2: THE PRINCIPLE OF IMPACT

Q. When we say ideas have consequences, what does that mean?

IT'S NOT POSSIBLE TO STRONGLY BELIEVE SOMETHING WITHOUT THAT STRONG BELIEF STRONGLY AFFECTING OUR LIVES.

Q. Should we even be talking about Creation? Isn't it divisive? Might we offend someone?

JESUS IS CALLED "THE ROCK OF OFFENSE" IN 1PETER 2:8. DOES THAT MEAN WE SHOULD WE STOP TALKING ABOUT JESUS? IF SOMEONE DENIES GOD IS OUR CREATOR, MAYBE THEIR SHALLOW THINKING NEEDS TO BE OFFENDED. THEY NEED TO HAVE THEIR IDEAS CHALLENGED.

Q. Darwin's theory is just a harmless idea, right? It doesn't affect our lives, does it?

KARL MARX, SIGMUND FREUD, ADOLPH HITLER AND FRIEDRICH NIETZSCHE AND OTHERS HAVE DEVASTATED OUR WORLD AND DARWIN WAS THEIR INSPIRATION. DARWIN THUS HELPED TO GIVE US HUMANISM, PERVERSION, COMMUNISM, RACISM, MASS MURDER AND MORAL BANKRUPTCY.

CHAPTER 3: THE PRINCIPLE OF AUTHORITY

Q. What is scientific law? What is theory? Does a scientific theory become a scientific law simply by repeating it over and over and drawing nice pictures?

SCIENTIFIC LAWS CAN BE DEMONSTRATED. A THEORY IS ONE POSSIBLE EXPLANATION FOR WHAT WE SEE. SIMPLY REPEATING SOMETHING OR DRAWING NICE PICTURES CAN'T MAKE IT A SCIENTIFIC LAW.

Q. What are the two different definitions of the one word "evolution?"

ONE DEFINITION IS THAT YOU LOOK DIFFERENT FROM YOUR PARENTS, AND THE OTHER DEFINITION IS THAT LIFE CAME FROM MUTATED MOLECULES, NOT GOD. THE FIRST DEFINITION IS SCIENCE, THE SECOND DEFINITION IS SILLY NONSENSE.

Q. What's the question you should ask of yourself when someone claims something that seems a little loopy like "The universe is nothing more than a lucky freak accident?"

"HOW IN THE WORLD DO THEY KNOW THAT?!!"

CHAPTER 4: THE PRINCIPLE OF TRUTH

Q. What happens when two ideas contradict each other? Are the Bible and evolutionary theory compatible intellectually?

LOGICALLY, CONTRADICTORY IDEAS CAN BOTH BE WRONG, BUT CANNOT BOTH RIGHT. THE BIBLE AND EVOLUTIONARY THEORY HAVE MAJOR CONFLICTS, AND WHILE BOTH CAN BE WRONG, BOTH CAN'T BE RIGHT.

Q. What is naturalism? Why is it intellectually odd for a scientist to be a naturalist?

NATURALISM IS THE BELIEF THAT NATURE IS ALL THERE IS. A NATURALIST IS LOOKING FOR LAWS THAT EXIST WITH NO LAW GIVER TO GIVE THEM, FOR DESIGN WITH NO DESIGNER. SEEMS NOT ONLY ODD BUT DERANGED.

Q. Is there absolute truth? Is evolutionary theory absolute? How about creation?

ABSOLUTELY, THERE IS ABSOLUTE TRUTH. EVOLUTION IS ABSOLUTE GUESSWORK. CREATIONIST THEORY MAY BE WRONG, BUT THE BIBLICAL ACCOUNT OF CREATION IS TOTALLY, ULTIMATELY, ABSOLUTELY TRUE.

CHAPTER 5: THE PRINCIPLE OF MEANING

Q. What does the Greek word LOGOS mean?

LOGOS MEANS: WORD, COMMUNICATION, LOGIC, REVELATION, INTELLIGENCE, REASON, PURPOSE, EXPLANATION... BASICALLY, IT MEANS MEANING.

Q. What is nihilism? Did you look it up? How is it related to evolutionary theory?

NIHILISM IS THE REJECTION OF ALL RELIGIOUS AND MORAL PRINCIPLES, BELIEVING THAT LIFE IS MEANINGLESS: A LOGICAL CONCLUSION OF DARWINIAN EVOLUTION. IT'S DEADLY. LEARN TO RECOGNIZE IT IN THE THINKING OF PEOPLE AROUND YOU, IN WHAT YOU READ AND HEAR.

Q. What does the Creator of the universe want from us? How does that work?

OUR CREATOR WANTS A RELATIONSHIP WITH US! HERE'S HOW IT WORKS: WE FOLLOW JESUS. WE TURN FROM OUR WILL AND BEGIN TO FOLLOW GOD'S WILL. WE ARE BORN AGAIN AS A NEW CREATION. THIS NEW LIFE HAS WONDERFUL MEANING BOTH NOW AND IN THE LIFE TO COME AFTER WE DIE.

APPENDIX 1: Q&A

APPENDIX 2: HOW TO BE BORN AGAIN

To be part of this world, we must be born. To be part of the world to come, we must be born again. It really is that simple. And it's urgent because there will come a time in each of our lives, which only God knows, when it will be late to escape the destruction of this world. Be born again! Surrender to the Lordship of Jesus Christ! The time to do that is today!

CREATION: WE BELONG TO GOD FOR ONE HUGE REASON – HE MADE US

Before He made time and space, God existed. He is not part of this universe. He always has been, always is, and always will be. As Creator, He is lawfully the owner and the master of everything. He is holy, eternal and completely other than His creation. Being our Creator is what makes Him our God.

"In the beginning God created the heavens and the earth." —Genesis 1:1

God created us for relationship with Him and with each other. God wanted us for His own family, enjoying the riches of His love, experiencing the wonder of His life.

"To those who believed in his name, he gave the right to become children of God." —John 1:12

"How great is the love the Father has lavished on us, that we should be called children of God! And that is what we are!" —1John 3:1

THE FALL: WE DISOBEYED GOD AND DID WHATEVER WE FELT LIKE DOING

It's easy to see that something's horribly wrong with our world and the people in it. That includes you and me. What's wrong is sin. Sin is falling short of what God expects from us, putting something or someone in his place.

Another way to say it is this: we know what's right but for some reason, we still do wrong. We choose to disobey God's rules. Because of our disobedience, we're in trouble: we are now separated from God and His life.

"All have sinned and fall short of the glory of God, and are justified freely by his grace through the redemption that came by Christ Jesus." —Romans 3:23-24

Because of our disobedience, this world has become a battlefield between the Kingdoms of light and darkness. The Kingdom of Light is where God's will is joyfully obeyed. The kingdom of darkness is where everyone selfishly does whatever they feel like is best for them.

Every day, whether we realize it or not, this war is the main thing that is going on all around us.

"And there was war in heaven. Michael and his angels fought against the dragon, and the dragon and his angels fought back. But he was not strong enough, and they lost their place in heaven. The great dragon was hurled down -- that ancient serpent called the devil, or Satan..."
—Revelation 12:7-9

REDEMPTION: GOD RESCUED US AT A HORRIFIC COST TO HIMSELF

Something must be done about me. I have to be made right with God again. My sin must be washed away once and for all. I desperately need a Savior.

Only one person in all the world is right with God, only one has never broken a single one of God's laws, only one person could possibly be the Savior of the world. That one person is Jesus Christ.

"Jesus answered, 'I am the way and the truth and the life. No one comes to the Father except through me.'" —John 14:6

God meets each of us with forgiveness and mercy; he would find us no matter how far he had to go.

You and I can't do a single thing to earn God's love and forgiveness: it's a gift God graciously grants each one of us. There are no magic words we can say to get us out of our trouble with sin. There's no pilgrim journey or impossible task we can perform to earn it. It is freely offered to us. We must receive it as a gift.

"For it is by grace you have been saved, through faith and this not from yourselves, it is the gift of God, not by works, so that no one can boast." —Ephesians 2:8-9

Jesus died on the cross and took the punishment for sin on Himself. We can't take any credit. We didn't have to do anything but agree and accept his gift of new life in Christ.

APPENDIX 2: HOW TO BE BORN AGAIN

"Jesus is the image of the invisible God, the firstborn over all creation. For by him all things were created: things in heaven and on earth, visible and invisible. For God was pleased to have all his fullness dwell in him, and through him to reconcile to himself all things, by making peace through his blood, shed on the cross."—Colossians 1:15-20

One person in the Bible (the very short Zacchaeus) was saved when he obeyed Jesus by simply coming out of a tree when Jesus told him to.

One person (the thief on the cross) was saved when he asked Jesus to remember him in heaven.

One person (Abraham) was saved by "faithing" in God. He trusted God and left everything behind in confident obedience when God called, later entrusting his most precious son Isaac into God's hands when asked.

One person (Moses) made God his best friend and as he spent time with God, he became part of God's family.

The 12 disciples heard Jesus say, "Follow me," and in following Him and doing what He said, they discovered that they were saved.

"Jesus went into Galilee, proclaiming the good news of God. 'The time has come. The kingdom of God is near. Repent and believe the good news!'" —Mark 1:14-15

The Kingdom of God is near. Did you know?

APPENDIX 2: HOW TO BE BORN AGAIN

To follow Jesus, I repent (turn from my sin) and believe the good news: Jesus rescues me as I trust and obey Him. I become a disciple of Jesus.

I stop being so concerned about what I want and what I expect. I begin to be more concerned about what Jesus wants, and what He expects.

"Jesus said to them all: 'If anyone would come after me, he must deny himself and take up his cross daily and follow me. For whoever wants to save his life will lose it, but whoever loses his life for me will save it.'" —Luke 9:23-24

When I turn from my own selfish will and surrender to Jesus, I "die" to my willful, rebellious nature. I don't die bI connect with God's grace that is constantly waiting to meet me. When I put my faith in Jesus, I am born from above and become part of God's family. For some, this happens over time; for others, it happens in the blink of an eye.

"Jesus declared, 'No one can see the kingdom of God unless he is born again. For God so loved the world that he gave his one and only Son, that whoever believes in him shall not perish but have eternal life.'" —John 3:3 and 16

There are truths that all Christians believe: that Jesus is the Christ, the Son of God. Do you believe that? And that He died on the cross to pay with His own lifeblood for the wrong things we do.

Do you believe that?

And, all Christians believe that Jesus rose from the grave, victorious over death. Do you believe that?

"[Jesus] asked, 'Who do you say I am?' Simon Peter answered, 'You are the Christ, the Son of the living God.' —Matthew 16:15-16

"I want to remind you of the gospel I preached to you, which you received and on which you have taken your stand. By this gospel you are saved ... For what I received I passed on to you as of first importance: that Christ died for our sins according to the Scriptures, that he was buried, that he was raised on the third day." —1Corinthians 15:1-4

Inside you somewhere, there is a guest bedroom. Some call it your spirit, some call it your heart. It's the deepest part of you. This is where you invite Jesus to bring all his luggage, come in, and stay forever, to be the Boss of your life. This is where you ask Him to set up His command center because you want to start doing things His way.

"Jesus replied, 'If anyone loves me, he will obey my teaching. My Father will love him, and we will come to him and make our home with him.'"—John 14:23

The only requirement in talking with God is that we are being real with Him. We can't possibly fool Him. Tell Him everything. Ask Him to save you. Ask Him to help you to follow Jesus from now on.

We can always just pray this prayer: "Yes, Lord!"

APPENDIX 2: HOW TO BE BORN AGAIN

Jesus commands us to repent and believe. Repent means to turn away from sin and turn toward God.

Believe means to trust that Jesus is exactly who he says he is.

Remember, your prayer can simply be *"Yes, Lord."* I wrote out a prayer if you want to use it. The important thing is to turn to God and say you are giving your life back to Jesus!

"JESUS, I SAY YES TO YOU. I REPENT, I TURN AWAY FROM BEING THE BOSS OF MY OWN LIFE. I NOW TURN TO YOU. BECAUSE I KNOW THAT YOU MADE ME, I RECOGNIZE THAT YOU KNOW ME WELL. I WILL DO WHATEVER YOU SAY. I WANT WHAT YOU WANT. THANK YOU FOR MAKING ME. THANK YOU FOR COMING TO EARTH AS THE BABY IN BETHLEHEM, DYING ON THE CROSS FOR THE SIN OF THE WORLD: FOR MY SIN. THANK YOU FOR RISING FROM THE DEAD, BRINGING ME ETERNAL LIFE. FORGIVE ME WHEN I DISOBEY YOU. I SURRENDER TO YOU AS MY KING. ALL OF ME NOW BELONGS TO YOU. I BELIEVE YOU ARE WHO YOU SAY YOU ARE. I AM YOURS TO DO WITH AS YOU PLEASE, WHERE YOU PLEASE, WITH WHOMEVER YOU PLEASE. I WANT TO GO WHERE YOU SEND ME, DO WHAT YOU TELL ME, SAY WHAT YOU GIVE ME TO SAY. I LOVE YOU BACK, LORD JESUS. MAKE ME PART OF YOUR FAMILY. MAKE ME A CHILD OF GOD."

RESTORATION: WE ARE RECONCILED TO GOD AND TO ONE ANOTHER

God made a way for us to have a relationship with Him because He loves us. God has big plans for us, here on earth and in the coming ages.

"Then I saw a new heaven and a new earth ... And I heard a loud voice from the throne saying, 'Now the dwelling of God is with men ... He will wipe every tear from their eyes. There will be no more death or mourning or crying or pain, for the old order of things has passed away. He who was seated on the throne said, 'I am making everything new!'"
—Revelation 21:1-5

When you are reborn from above, you're brand new person. Tell your dad and mom and family what happened to you. Tell your teachers, your friends. Tell everyone you know.

Just tell them what really happened and don't dress it up. That's called being a "witness."

"You will receive power when the Holy Spirit comes on you; and you will be my witnesses in Jerusalem, and in all Judea and Samaria, and to the ends of the earth." —Acts 1:8

Form a habit of reading the Bible. Ask the Holy Spirit to tell you what it means. The Gospel of John is a great place to start.

Maybe keep a journal of your new life in Jesus. Go back and read it to step back and see the big picture! One thing you'll discover when reading the Bible is that believers in Jesus were nearly always baptized immediately if at all possible.

APPENDIX 2: HOW TO BE BORN AGAIN

As the Ethiopian said to Philip, "Look, there's water: is there any reason I should NOT be baptized?!" So get baptized.

Hang out with others who believe in Jesus. Church is a great place for that.

If you want to know what Christians believe, read the Apostles' or Nicene Creeds.

Ask God to show you if there's anyone you have not forgiven and immediately, by faith, tell God you forgive them and mean it. If they don't know they hurt you, don't walk up and blab that you forgive them; just show it by what you do. Ask for any forgiveness the Holy Spirit points out to you.

"For if you forgive men when they sin against you, your heavenly Father will also forgive you. But if you do not forgive men their sins, your Father will not forgive your sins." —Matthew 6: 14-15

Above all, keep following Jesus. Whenever you get discouraged don't stay that way, when you get down, don't stay down. Grab hold of Jesus's outstretched hand and let him help you get back up. He shares his victory with us. Don't give up, ever.

Because God has not held back anything he has to give. That's why we give him back everything we have to give.

"If God is for us, who can be against us? 32 He who did not spare his own Son, but gave him up for us all how will he not also, along with him, graciously give us all things?
— Romans 8:31-32

APPENDIX 2: HOW TO BE BORN AGAIN

When life gets hard, remember that Jesus is not only beside you, He's inside you, always. Tell Him everything all the time. Hang on to Him. Talk with Him. He's your best friend.

What do you do with your best friend? Ignore them? Or talk to them all the time about anything at all? Can we talk to Jesus and hear him? Jesus seems to think so. Check out the following verse.

"My sheep listen to my voice; I know them, and they follow me. I give them eternal life, and they shall never perish; no one can snatch them out of my hand. My Father, who has given them to me, is greater than all; no one can snatch them out of my Fathers hand." —John 10:27-29

Obey Jesus. Serve Him by serving others. There are "good works," things we will do that will last forever.

"For we are God's workmanship, created in Christ Jesus to do good works, which God prepared in advance for us to do." —Ephesians 2:10

As great as salvation is, it only gets us back to the starting place. God has much more planned for us than just being good! Our lives have now become God's training ground to prepare us for living forever in our brand new, indestructible bodies.

"Jesus said to her, 'I am the resurrection and the life. He who believes in me will live, even though he dies; and whoever lives and believes in me will never die. Do you believe this?'" —John 11:25-26

APPENDIX 2: HOW TO BE BORN AGAIN

Here's what salvation does for us: it dramatically affects our past, our present and definitely our future!

- We *were saved* at the cross where Jesus paid our sin debt with the sacrifice of his sinless life.
- We *were saved* when we first trusted in Jesus.
- We *are being saved* as we grow in Jesus day by day by day by day.
- We *will be saved* when we rise to life after we die.
- And in the endless ages to come, *we'll be saved* forever beyond our wildest dreams.

"And God raised us up with Christ and seated us with him in the heavenly realms in Christ Jesus, in order that in the coming ages he might show the incomparable riches of his grace, expressed in his kindness to us in Christ Jesus."
—Ephesians 2:6-7

We were helpless and hopeless until Jesus appeared.

"At one time we too were foolish, disobedient, deceived and enslaved by all kinds of passions and pleasures. We lived in malice and envy, being hated and hating one another. But when the kindness and love of God our Savior appeared, he saved us, not because of righteous things we had done, but because of his mercy. He saved us through the washing of rebirth and renewal by the Holy Spirit, whom he poured out on us generously through Jesus Christ our Savior, so that, having been justified by his grace, we might become heirs having the hope of eternal life. This is a trustworthy saying. And I want you to stress these things, so that those who have trusted in God may be careful to devote themselves to doing what is good." —Titus 3:3-8a

APPENDIX 2: HOW TO BE BORN AGAIN

Salvation goes beyond being saved from sin and death and hell. It's how God does everything he does!

Salvation is God's trademark. It's how he does everything. He lives to continually save us! Praise His holy name for so great a salvation!

"The LORD is my strength and my song; he has become my salvation. He is my God, and I will praise him, my fathers' God, and I will exalt him."—Exodus 15:2

"I looked and there before me was a great multitude that no one could count, from every nation, tribe, people and language, standing before the throne and in front of the Lamb. They were wearing white robes and were holding palm branches in their hands. And they cried out in a loud voice: 'Salvation belongs to our God, who sits on the throne, and to the Lamb.'"—Revelation 7:9-10

"Praise be to the God and Father of our Lord Jesus Christ! In his great mercy he has given us new birth into a living hope through the resurrection of Jesus Christ from the dead, and into an inheritance that can never perish, spoil or fade kept in heaven for you, who through faith are shielded by Gods power until the coming of the salvation that is ready to be revealed in the last time. In this you greatly rejoice, though now for a little while you may have had to suffer grief in all kinds of trials." —1Peter 1:3-6

"For the Son of Man came to seek and to save what was lost."
— Jesus, defining his reason for coming to Bethlehem, living a perfect life and dying for the sin of the world in Luke 19:10.

APPENDIX 3: KINETIC ACTIVITIES

Chapter 1: Hide candy bars in the lobby, one per kid, and bring the kids to the lobby to search. Let them know that the world is stuffed full of God's fingerprints, waiting for them to discover. Talk about what those fingerprints might be. Tie in "curiosity."

Chapter 2: Bake two cakes but don't let the kids see them. Bring one out, and talk in the abstract about eating cake, then throw the cake away. Tell them it's no big deal whether they eat the cake or not. When the kids protest, ask why they care. So what? Bring out the second cake and talk about how important it is to take things seriously while you share it. Talk about the impact of decisions.

Chapter 3: Tell the kids you want to show them "a rock" but show them a map of "Iraq." Talk about the deception involved in the two meanings of the word "evolution." Practice saying the phrase, "How in the world do they know that?" Talk about authority, whether we should believe what a scientist might say or what the Bible says. Make sure they know you honor scientists, but that scientists are just people like us who make mistakes and don't know everything!

Chapter 4: Take a jigsaw puzzle. Talk about truth as you sincerely put the pieces in any way but the right way. Be frustrated that you can't see the "big picture." Let the kids put the pieces in correctly and point out that, no matter how sincere someone may be, their ideas are either right or they are wrong. There is an "absolute right" way the pieces fit!

Chapter 5: Take multicolored handfuls of gooey and sticky dough. Throw it randomly into a shallow plastic tub, drop it, mangle it and mash it down. Ask the kids what the gooey sticky thing MEANS. Talk to them about meaning and how it can't come from random processes, but only from a Creator. Random processes cannot generate meaning.

Final Activity: Have a party with ice cream, cake and punch. Show part of the movie called The March of the Penguins, or invite a doctor to come and describe the heart, or the brain, or the birth of a baby or the flight of a hummingbird in the light of the marvelous purposes and design skills of the Creator.

APPENDIX 4: DIGGING DEEPER

This section is for those of you who want to think more about this subject. It's hard work to dig into a section like this, but I'm betting you like a challenge. I have not attempted to correlate the subject matter in this section with the five chapters of the book, the reason being that the amount of information available on this subject is so vast. It just won't fit. So I've inserted topical headings to give you an idea of what's being discussed on each page.

A good way to start is by flipping through and simply glancing at the headings.

Think of this section as a grocery store. Roll your mental shopping cart down the aisles and see what interests you. Try out new ideas. Different people think different ideas are crucial to this discussion of Biblical Creation and Naturalistic Evolution. If your favorite topic isn't covered, blame limited space and attention span. The book needed to stop somewhere!

If this book belongs to you, take your pencil and mark in the margins, making note of things that grab you, or subjects you want to study further.

Explore and think new thoughts. Use the brain God made and gave to you!

Take your curiosity for a joy ride.

WE PRAY AND GOD REVEALS

I've been traveling and speaking to groups about Creation and Evolution for nearly two decades.

I've tried dozens of approaches, and none of them can convince someone who is uninterested in the truth.

If you have a friend who believes that all life came from protein soup in an unknown time and place, the first step towards reaching them involves finding out if they even WANT to know what actually happened.

Many of those with whom we have long discussions don't get it because they don't WANT to. They may say they are interested in truth, but they are being dishonest with you and possibly with themselves. Prayer is the only way of reaching them.

But prayer is never a last resort, it's a powerful weapon we should always rely on as we begin and all the way through our discussions! According to 2Corinthians 10, **"For the tearing down of strongholds, throwing down imaginations and every high thing that is exalted against the knowledge of God, and bringing every thought into captivity to the obedience of Christ."** (WEB)

If you have a friend who will not see, that's the signal to pray. Pray and never give up. Keep on praying. Weeks? Months? Years? It doesn't matter. Pray.

Perhaps God will one day touch their heart with truth. We can't argue anyone into the truth.

But we can win people through persistent prayer and words the Holy Spirit gives us.

INTELLECTUAL HONESTY

My pastor Ron Ferguson tells a true story about a man who, in a counseling session, told his minister he didn't believe in God.

"How much of the total knowledge of the universe do you possess?" asked the minister. "I know probably half of everything there is to know," responded the man. "That's a lot of the world's knowledge for one man," remarked the minister dryly, "Could God exist in the half of the knowledge of the universe you DON'T possess?" "I suppose he could," was the reply.

The minister asked the man a very strange question, "Are you an honest doubter or a dishonest doubter?"

Surprised, the man asked the minister what he meant. "If God revealed himself to you, would you believe in Him?" "Yes, I would," the man answered.

"Then," the minister challenged, "read the gospel of John, and come back and tell me what you find."

The man did. He gave his life to Jesus.

The minister asked a profound question. Honest doubter or dishonest doubter? Not all who doubt God do so for honest intellectual reasons. Some are using their doubt as a smoke screen for something else that's seriously wrong in their life. Usually, people who are dishonest doubters are practicing sin and do not want to deal with a Creator who might disapprove.

If anyone genuinely wants to know the truth, they don't have to find the truth, the truth will find them!

APPENDIX 4: DIGGING DEEPER

WHAT ARE THE UNDERLYING ASSUMPTIONS

It's important for us to know what "science" is before we declare something to be either scientific or unscientific. Right? Because until we know the definition of, say "bourgeois," we can't really say if an idea is bourgeois or not.

So what is science? According to my freshman geology textbook, "Science is a way of looking at things." Really? Science is a way of looking at things? Because that means science is a synonym for opinion or perspective and surely that's not what science is. The writers of my textbook needed to jump right into a chapter on evolution, and so the definition of science, which would exclude evolution, needed to be changed to accommodate a naturalist ideology.

But science is not simply a way of looking at things. It is a process that includes formulating theories that must be **demonstrable**, **reproducible**, **falsifiable** and **predictive**. At least, that's what science has been! This is the science that is useful to mankind, called Baconian science after Francis Bacon. It has expanded our world in wonderful ways.

Demonstrable means you can show it happening now. Reproducible means it's a repeatable demonstration. Falsifiable means that there would be outcomes of the theory that would make you scrap the theory. And predictive means that you can see a little bit of where we're going by applying your theory to the world around us.

Evolution is not science. It cannot be demonstrated, repeated, falsified or predict anything. It is simply a story made up by people who don't want a Creator having sovereignty over their lives and attitudes and behavior. The explanation of Creation has much more scientific foundation than evolution has ever had, though you wouldn't know it from the scientific world we presently live in!

WORDS MEAN THINGS

Evolution #1: **variation within a species**, using the rich genetic information provided by the Creator.

Evolution #2: **transmutation into brand new species** from arbitrarily defined "lower" forms of life (smaller? simpler? heavier? uglier?) into arbitrarily defined "higher" forms (larger? more arms? more thumbs?)

Those who claim that #1 proves #2 are not **thinking**. Look. If evolution merely means that I look different from my parents, then I believe in evolution with all my heart. I've got no beef with that definition of evolution.

BUT... If evolution means hydrogen gas became my great grandfather through time because of explosions and cosmic radiation, then I think I'd have to be either nuts or overeducated to believe something so obviously kooky and deranged.

Evolution number one doesn't prove evolution number two. Not even remotely. Evolution number one is science. Evolution two is unproven and unprovable hogwash.

Again, animal husbandry and botanical hybrids are indisputable, experimentally repeatable, everyday science. But fish crawling up on land after mutating into air breathing land animals is comical, small-minded, and never-could-possibly-happen-under-any-circumstances insanity. Zero probability.

Be an informed thinker. Before getting too far into a discussion about evolution, ask the question, "Are we talking about scientifically demonstrable variation **within** a species, or mythical **transmutation** from rocks into people over time?"

SPECIES ARRIVE DISTINCT, AND LEAVE EXTINCT

Don't just blow past the distinction between the two uses of the word "evolution." Remember, one is variation *within* a species and the other is transmutation *into a whole new species*. Once people understand this incredibly dishonest misuse of the English language, there will no longer be any debate between creation and evolution.

In Genesis 1:11, 1:12, 1:21, 1:24, 1:25, 6:20 and 7:14, there is a critical word used in the creation of life the Holy Spirit wants us to know. He repeats it seven times. It's the word "**kind**." "Kind" is the same as species. Dogs have puppies, seals have whelps, chickens have chicks, the Amish have omelettes (Just KIDDING -- they have children) after their *kinds*.

God introduced the whole species thing. Hybrid corn has been cross pollinated, dogs of different breeds can have mutts, people of different races have kids all the time. But two creatures of different species (kinds) *absolutely cannot reproduce*. It's the definition of the word "species."

The fossil record backs up the Biblical usage of the word "kind." There is no fuzzy transition, nor smooth morphing from one species into another, species arrive in the fossil record complete, intact, fully functioning, beautifully designed. The fossil record proves God made everything.

The pretty evolutionary charts textbooks draw have nothing to do with any actual fossil data! When you look at the rocks, the fossils are all in "kinds," none of them are transitions or transmutants. The geologic epochs in textbooks come from Darwin's theory, not the fossil record. There is *not one single representative* of transition to a new species in the fossil record. Not one.

It's indisputable, species are immutable.

WE THINK WE'RE SMARTER THAN GOD!

In the 1700s, Western Civilization underwent a transformation called The Enlightenment. The "Dark Ages" were superseded by the "Age of Reason." The mind was exalted. Faith in God was labeled backwoods superstition. Mankind was making a giant leap forward from shadow to illumination by hitching his wagon to his impressive powers of thought.

If the devil can get us looking at the gift of our intellect and not the Giver, it makes his day. The kingdom of darkness had a slick strategy: to get man irrationally enamored with the mind God gave him while deleting thoughts of God, the Giver of that mind.

A century after the rise of "reason," Darwin proposed his theory, and the Age of Reason had it's weapon to bury God. It would call the weapon "science," but there was not anything scientific about Darwin's theory. Science is observable, repeatable and falsifiable and Darwin was none of those. They *said* it was scientific, and the battle was on.

God gets to make the rules because he made the people. But some don't like God's rules. What if "science" claimed God DID NOT make the people? Then the rules of God are gone with God himself! It would be party time! No rules! Finally a clean shot at God. Maybe he could be taken out.

But remember Emerson's old adage: if you strike the king, better be sure you kill him. Fat chance! God is doing just fine these days. Darwin vs God? Spit wads at a battleship!

Only 19% of our fellow humans, according to Gallup, believe in evolution, after all the evolutionary pictures and all the evolutionary textbooks and all the evolutionary indoctrination. Darwin's in trouble!

FOUR GUYS TRYING TO MAKE A FOOL OF YOU

Evolution is nothing more than repackaged atheism. Let's prove that.

In the late 1800s, *four men came to the forefront of The Enlightenment*: **Charles Darwin, Friedrich Nietzsche, Sigmund Freud, and Karl Marx.** They championed destructive ideas that have gutted whole nations.

Darwin gave atheism a scientific mask. Marx gave brutal dictators a populist face appealing to a pipe-dream called Utopia. Freud gave perversion an excuse. Nietzsche gave unbelief a philosophical justification.

All four abandoned God's revealed truth in Scripture to advance their crippling brand of nothingness. When these men had finished their work, the naturalists believed they ruled the world. An assault on rationality was unleashed.

But we cannot remove God from our thinking and have our thinking still be worth a flip. We can't take God out of anything without introducing death of some kind: intellectual death, moral death, spiritual death, physical death. Death has been the unmistakable fruit of the godless philosophies of these enlightened men. Hitler is only one example of the horror. Racism of every stripe is justified by Darwin. Death camps are not far removed from the idea of Survival of the Fittest.

The first chapter of Romans paints the retrogression that begins, not with behavioral sin, but *intellectual* sin: **"They knew God but did not glorify Him."**

Nietzsche arrogantly proclaimed, "God is dead." Today, we know God's feeling just fine, but Nietzsche's still ... well. He's still *dead*. Not likely to recover.

"I CAN'T SEE," HE SHOUTED WITH CLOSED EYES

"What is truth?" These haughty and cynical words came from Pontius Pilate, while he was passing judgment on Jesus, as Jesus stood directly in front of him. It's so ironic to imagine this pompous, self-important ruler asking Truth Incarnate, Jesus Christ, whether he *exists*.

Pilate did not know that the Man he was judging would one day be *his* Judge. In his self-destructive pride he mocked the existence of truth, of good and evil, right and wrong, of anything absolute. There are the physically blind and then there are those who simply refuse to see.

Today's scientific community, following suit, has ceased to inquire. The Darwinian naturalists mock religion as closed-minded and narrow and self-assured. Yet they do not realize how exactly like a closed, narrow, prideful religion their own pronouncements are!

Modern science has built a house of cards on a few wrong-headed ideas from 150 years ago and rejected all attempts to restore an intellectually honest foundation. Carl Sagan taught that the universe itself has eternally existed, but God? Someone invented him, he had to come from somewhere. Creation is eternal? God is created? Huh?

C.S. Lewis wrote that ancient man saw God correctly as the Judge of our actions. Modern man has tried to reverse that, saying that we are now the judge of God and his actions, *elevating our intellect above God*. Just like Pilate, evolutionists look down at God from their own intellectual high horses and condemn *Him*.

It's a hellaciously dangerous position to occupy.

PERPETRATE FRAUD: CONTROL THE RAW DATA

When I was growing up, I was so sad for the Russians. They had to read something called 'propaganda.' They were only told what the Russian government wanted them to know. They could not think for themselves.

In this country, Hurricane Katrina became legend, but not for the truth: the media reported things about the storm which later on were discovered to be lies: murders, gangs, rapes, levees blown up. Never happened. None of it true. All of it fabricated.

Darwinian evolution stands or falls on the fossil record. Have you ever seen evolution happening in the fossil record? I never have and I've done postgraduate work in geology. I HAVE seen pretty charts with pretty colors that show the entire panorama of plant and animal life describing all the imaginings we have about it.

But no raw data. No information on the fossils beyond the stories. All the thinking has been done for us by really smart people. We couldn't be expected to examine the data and think for ourselves, could we?

The Darwinists mock the Dark Ages when only the priests were allowed to read the holy book in order to control the rest of the population.

Today, the priests of evolution never let the common people see the raw fossil record. No one is allowed to see the fossil proofs of frog to dog evolution.

And why do you think that is?

Because there aren't any proofs...

THE ATHEISM OF EVOLUTION

I'm always amazed at people who don't believe that Darwinian evolutionists are against God. They'd be surprised at statements like the one below:

1995 POSITION STATEMENT OF THE AMERICAN NATIONAL ASSOCIATION OF BIOLOGY TEACHERS
"The diversity of life on earth is the outcome of evolution: an *unsupervised, impersonal*, unpredictable and natural process of temporal descent with genetic modification that is affected by natural selection, chance, historical contingencies and changing environments." Italics are mine.

This statement is quoted from a book by Phillip Johnson: DEFEATING DARWINISM BY OPENING MINDS, who notes that the words "unsupervised and impersonal" were removed from the statement after the original publication of his book! Even in it's sanitized version, the statement essentially calls the Bible a book of myths and labels God a liar.

Sir Isaac Newton is perhaps the greatest of all scientists. In his book, Mathematical Principles of Natural Philosophy Newton stated, *"This most beautiful system of the sun, planets, and comets could only proceed from the counsel and dominion of an intelligent and powerful Being. This Being governs all things, not as the soul of the world, but as Lord over all."*

Newton says that naturalist scientists are on the wrong side of the truth.

Evolutionists would tell him to just shut up and go stand in the corner with the crazy Christians. Which would be fine with Newton who was a devout Christian himself!

APPENDIX 4: DIGGING DEEPER

THE RELIGIOUS DOGMA OF EVOLUTION

Evolution is believed in spite of evidence that says it couldn't possibly have happened *(such as the existence of mind, the order of the universe, the existence of the language of DNA)*, and is perpetuated by indoctrination *(public schools)*; having its own priests *(naturalist scientists)*, sacraments *(endless TV documentaries perpetuating evolution without proving it)*, sacred books *(THE ORIGIN OF SPECIES and virtually every textbook since 1963)*, cathedrals *(Natural History Museums, forget that, museums in general)*, hallowed ground *(Olduvai Gorge)*, and sacred relics *(the fossils of* Archaeopteryx*, and Eohippus)*.

If you disagree with an evolutionist, what follows is not an intellectual discussion, but a nuclear reaction, beginning with, "Oh you must be a knuckle-dragging, Bible-thumping, backwards-thinking fundamentalist." The evolutionist will implore you not to take the world backwards to the time before evolution was proven, even though no proof has ever been offered at any time in 150 years. If intimidation or using obscure terminology won't work, your intellect is the next thing to be attacked: it goes on the chopping block.

If evolutionary theory is as scientific as claimed, it should be up to a little gentle questioning. That evolution is shallow dogma is easily seen by the manner in which evolutionists bristle at the mere request for intellectual inquiry. Only a religion, based on faith, with nothing to be proven is as gently treated as evolutionary theory. Evolution cannot stand up to rough handling, but must be coddled and protected by shaming anyone who dares question it's validity.

Evolutionists say their theory is scientific law: as settled as the law of gravity. Why does no one ever claim that gravity is as settled as evolution?

CARL SAGAN: CLEVER TOOL OF NOTHINGNESS

Carl Sagan told us that the universe is infinite, ageless and all-powerful *and that God was created*. The Smithsonian lionizes Sagan as "irreplaceable." We should look at what he said and what he did: it must have been profound and brilliant.

Carl Sagan said the following, "The God theory might be right — It's certainly possible, particularly a God who is sufficiently remote in the scheme of causality, who does not intervene in everyday life — it is perfectly possible that something called God, I don't know exactly what it is, set things in motion ..." He thinks it's okay if you want to believe in God, you have his permission to do so.

But science, especially the teaching of evolution as science, according to Sagan, is the way to *control* people! He says the following: "And the idea that [science] is only for a small elite or a privileged few who have the time and inclination and, by the way, money to spend ten or fifteen years preparing, is simply wrong. It is the birthright of every human being to understand science. And it's essential if we are to control, particularly in a democracy, a future which depends so clearly on science and technology."

Is Carl Sagan smart? Sure. Was he on TV? Oh yes. He was a great salesman for the two most flawed beliefs in the Naturalist view of the world: God is irrelevant and the evolutionary theory of life origin is science.

Sagan represents the modern "scientific" mind. He has deceived a generation.

I cry for those people.

WHY SMART PEOPLE FALL FOR EVOLUTION

The true mindless believer is not the Christian, but the one who accepts evolution as settled.

I'll never forget presenting some ideas about the emptiness of evolution to my brother Dave, a botanist and a brilliant dentist. His response is worth noting: "I'd never stopped to wonder if evolution was true. Everyone I knew at school just accepted it like I did."

Evolutionists love to mock Christianity, trash talking Christians for believing things by faith alone.

But consider the blind faith of the evolutionist in the unseen and the unprovable: *Evolution is the story of unknown substances in an unknown past and an unknown place under unknown conditions producing unknown organisms through unknown processes over an unknown time.*

As Carl Sagan said, evolution is "a still not fully worked-out but scientifically supported and beautiful scheme." Not fully worked out! It's only missing all the steps! There is not a episode of evolution that makes any sense at all!

People who go for years of education gradually begin to believe what they read in a textbook more than their "own lyin' eyes." They are conditioned to do so. They miss the obvious and fall for the devious. And evolution is everywhere present in higher education. It is unquestioned and supreme among ideas. What fanatical faith and mindless devotion the Darwinian naturalist has among the "smart set!"

Dr. Henry Morris put it this way: "The reason most educated people believe in evolution is that they've been told that most educated people believe in evolution."

FACTS ONLY CONFUSE AN EVOLUTIONIST

An evolutionist gives up a lot to maintain his or her atheism, not the least of which is their ability to reason logically. I will demonstrate.

The winner of the 1967 Nobel Peace Prize in Science, Dr. George Wald, wrote, "When it comes to the origin of life on earth there are only two possibilities: creation or spontaneous generation [evolution]. There is no third way. Spontaneous generation was disproved 100 years ago, but that leads us only to one other conclusion: supernatural creation. We cannot accept that on philosophical grounds; therefore we choose to believe the impossible: that life arose spontaneously by chance."

That was a moment of honesty that came very close to the finding of truth! Only his long education kept him from seeing that evolution is impossibly flawed and not worth his belief.

Wald showed he was a not-so-brilliant analytical reasoner when he said, "One only has to consider the magnitude of the task to concede that [evolution] is impossible. Yet here we are as a result of evolution."

So Dr. Wald is not at all bothered by believing in the impossible. Which is just fine, except he doesn't allow the rest of us to believe in God who alone can DO the impossible. Why in the world would such a man mock those of us that wish to put our faith in God when, by his own admission, his own faith is so poorly placed?

Wald represents the mainstream naturalist. We are reminded of Paul's letter to the Romans in the first chapter: "Claiming to be wise, they became fools!"

SOME THOUGHTS ABOUT EXPLOSIONS

Astronomers have suggested that the universe began with a Big Bang, where all the matter of the universe (wherever it came from) was compacted in a tiny event horizon and rapidly expanded producing space and time.

In other words, it began by exploding.

Well, excuse me. I've seen explosions. They are NOT creative, but rather destructive. They don't organize matter, they *disorganize* it. They don't generate complex systems, but they reduce complex systems to quivering chaos.

Maybe I've been hanging around the wrong explosions.

Physics has immutable laws. The second law of Thermodynamics says: left to itself all matter becomes disorganized over time. The fancy name: entropy. An example would be your bedroom. You try to keep it picked up, but left to its own, it becomes less organized over time!

An explosion follows the second law of Thermodynamics quickly: in milliseconds, everything becomes UN organized!

And! Contrary to the immutable law of Physics, Darwinian evolution claims to accomplish the opposite of the second law of Thermodynamics. It declares that over time, with no outside influence, matter has become MORE organized into higher and more complex and efficient systems.

I charge evolution with a felonious rules violation: neglecting to check with the established laws of Physics. Once the facade is peeled away, the naturalist's religion crumbles like a toothpick castle.

The second Law of Thermodynamics wins, evolution loses.

THE END OF THE WORLD! CLAIMS THE *SCIENTIST*

Street preachers are laughed at these days as they proclaim the end of the world. People used to know God is going to end the world in judgement, now unbelief is mainstream.

Evolution *used* to be laughed at, now it's very fashionable. How that can happen? Let's look at Global Warming.

Not more than 30 years ago, scientists were warning us the earth was heading for another Ice Age. There were, and I'm not kidding, scientists who said we had to paint the ice caps black so they would absorb more solar heat and melt before we all froze to death.

That was then, this is now. The sky is still falling, but the apocalypse has a new story line. "Chicken Little" scientists, led by the prophetic meteorologist Al Gore and the World Council of Churches (with the help of misguided pastors) are *now* prophesying an overcooked planet. Rejoice! We're not going to freeze after all. Nope. We're still going to die horribly, but we'll be hot, not cold. Same data. New interpretation. No apology.

Naturalists openly laugh at Christians on street corners who cry out that the world is ending because of our sins. Then they turn around and bully others into believing their real sins are emitting carbon dioxide. The stuff all plants breathe!

And evolution? Have we found new data that supports Darwin? Not a chance. Evolution is just as silly and outrageous to rational thinking as ever. And it's still taught in every school and university on the planet.

One of my favorite quotes about evolution is the following from George Orwell: "Some things are so stupid, only an intellectual could believe them."

THE HEAVENS ARE TELLING THE GLORY OF GOD

The orbit of the moon is one of the most fantastically absurd things imaginable.

The moon's mass, velocity and direction must precisely match the gravitational pull of earth's mass, or the moon wouldn't be up there. What accounts for the orbit of the moon? Or more exactly, what accounts for the *speed* of the moon?

Silence... Because nothing can account for the speed of the moon. It's just freakishly wheeling along up there, mocking the mockers of God's design.

Look everywhere in the heavens. A celestial line dance has been going on for six thousand years or more. All the constellations are included in the dance, their steps are choreographed to perfection. Our whole universe is meticulously choreographed: spinning, weaving in precisely ordered movements with grace and accuracy. The breathtaking clockwork of the skies has nothing random, with pinpoint timing and dependability.

This could only be due to the ingenuity and power of Almighty God. And even though God's handiwork is inarguably obvious, still, whole university faculties with tons of scholarly degrees have written scads of words about how the cosmos is merely an accident of exploding gas. I can show you libraries of books that chart the sequence of events of exploding, cooling and the orbits produced by arbitrary gravitational pull alone. I'm not joking.

They would look at a Swiss watch and pronounce that there was a meteor that hit an iron mine and out popped this Rolex. Shazam.

STEPHEN HAWKING AND LEONARD MLODINOW

Hawking and Mlodinow came up with a theory that proposes as many universes as there are possibilities. We, by coincidence, happen to exist in the one universe where all the possibilities allow the universe to exist.

You think that this paragraph was hard to read, you should try constructing it sometime!

Anyway, since all the possible laws of the universe just happen to be workable in ours, these impersonal natural laws like gravity and electromagnetism squeezed out the entire landscape of life unintentionally. Quite a theory.

These two men are considered brilliant. They are able to understand and discuss theoretical ideas you and I could not articulate in our wildest dreams. And yet, when confronted with an obviously created cosmos, the first thing they do is throw the idea of a Creator out of it. Why? Is that smart or stupid? Brilliant or lame-brained?

The Creator can easily handle all the questions concerning the existence of His cosmos, but that's not acceptable to these great intellects, so they imagine laws of a universe existing before the universe whose functions these laws proscribe. Again, it's hard to write their reasoning in reasonable words. They believe in laws that exist without a Law Giver, and proclaim those random laws have worked the very miracles they don't believe in.

They think they can kick God out of the tiny minds He gave them to think with. Trying to be smarter than God is like trying to be older than your parents. It's like leaning over, grabbing your feet, and trying to lift yourself off the ground.

LOGIC IS NOT A DARWINIAN NATURALIST'S FRIEND

"It is not possible for God to not exist." Ravi Zacharias

There is a Latin proverb that says, Ex Nihilo, Nihil Fit. Don't ask me to pronounce that. Latin is a dead language and nobody speaks it. But do you recognize the word "nihil?" That's right, it has the same root as the word "nihilism." And the same root as nil (zero) and annihilate (destroy). You are one sharp cookie.

Nihil means "nothing." Ex Nihilo means "from nothing." Nihil Fit means "nothing comes." It is not possible to get something without having something first. From nothing comes nothing. From nothing, nothing can come.

In other words, since the universe is *something*, it did not come from *nothing*. Logic.

Julie Andrews playing Maria Von Trapp in the movie "The Sound of Music" sang: *"Nothing comes from nothing — nothing ever could."* Do you understand that Julie Andrews is making more sense than all the learned scholars who have ever believed and taught Darwinian evolution? She's absolutely right: it's not an option for something to come from nothing, since nothing comes from nothing. I think she said it better. She definitely sang it better.

Ravi Zacharias was asked by a smart student, "How can I know that I exist?" Without hesitation, Dr. Zacharias looked the student in the eye and replied, "Whom shall I say is asking?" The lecture hall broke into laughter!

If there IS something (and I'm sorry but there is), then it *came* from something. It's the Law of Causality from chapter five.

IF A PROPHET WAS WRONG, YOU STONED HIM

I can't tell you the number of times I've read scientists saying they discovered some fossil or outcome that they "weren't expecting." It "surprised" them and "changed everything they previously thought."

But that never humbles them, they just brazenly assert their new opinion as something they'll never be forced to retract later, announcing that the thing they weren't expecting changes everything they had previously thought including the last thing they asserted from the thing they were surprised about before.

In other words, scientists get infinity do-overs. Is that fair?

In the Old Testament, the law said you could stone a prophet if he told you something wrong. These days, if one of our scientific "prophets" is wrong, we give them a mulligan and breathlessly await their next pronouncement. I wonder if the Global Warming alarmists would let us stone them if the world doesn't cook in 10 years? Wait, it's been 10 years and it's as cold as all get out.

How about naturalist scientists who call God a liar and the Bible a myth? Are they willing to bet their lives? Eternally speaking, I guess they are! I guess they are betting the farm on there being no Judge and no Judgment Day. Their track record on being right tells me to run away! These scientists are as wrong as the weatherman, and for some reason are still accorded attention and authority with every word that falls from their lips. I love science! I really do! But I don't put my faith in scientists!

"They don't know half of what they think they know if they're continually surprised at what they learn."
– Rush Limbaugh

NO EVIDENCE? BLUSTER AND INTIMIDATE

Those who write our science textbooks have no interest in truth. I don't say that lightly. You would think they would care about truth, but in reality, their interest is in making sure the children are not harmed by evil "creationism."

The real "ism" we are dealing with is *Darwinism*. Evolution is unproven and unprovable, so much so that in the absence of any proof, those questioning this "ism" must be mocked and shamed instead of shown reproducible science.

Science is the god of *naturalism* (that word *is* in the dictionary). They believe science contains all knowledge. In our Creator's place, they worship the unprovable and destructive theory of evolution that's easily refuted by probability, logic, the fossil record and plain common sense. Yes and science! They hammer evolution into young and unsuspecting student brains as a religion without evidence.

The American Institute of Biological Sciences boasts: *"The theory of evolution is the only scientifically defensible explanation for the origin of life and development of species."* They never present the scientific defense, they just assume everyone knows it.

The American Association for the Advancement of Science (yeah, right) piles on: *"The evidences in favor of evolution are sufficient to convince every scientist of note in the world."*

Bombast! Now, don't be intimidated or anything, but if you aren't convinced that evolution happened, you must be a lower class of scientist. You're certainly not, according to the AAAS a "scientist of *note*!" Somehow, every scientist of note is <u>not</u> convinced. One of those scientists said this: *"I want to know how God created this world."* His name? Albert Einstein.

WE LET THE DARWINISTS IN, THEY KICK US OUT

"For God's sake, let the children have their minds kept open. Close no doors to their knowledge... Make a distinction between theology and science. Let them have both. Let them be taught. Let them both live." So begged Clarence Darrow in the 1925 Scopes "monkey trial," pleading for evolution to be allowed to be taught. In the interest of fairness, teach evolution along with the Bible.

The first words of this quote from the lawyer arguing that evolution should be taught in schools are "For God's sake!" I'm speechless. For God's sake? God's sake was the LAST thing he was concerned with. Clarence Darrow's usage of that particular phrase was comical!

But the irony is painful. Eighty-one years ago, the Darwinists were pleading for fairness. Not, they said, that religion is wrong, or that the Bible is a myth, or that any child should not be taught about creation. The Darwinists argued that BOTH creation AND evolution should be taught in order to "close no door" to the minds of the children.

As I traveled the US telling kids that God loves them, I was warned about singing anything that had to do with creation for the ACLU had spies who would swarm the school and sue the Principal for allowing *superstitious creationism*in the door. The ACLU is the muscular, hairy arm of science, ready to sue the pants off anyone who even breathes the name "Jesus" in a school setting; not even saying anything *about* Jesus, just his name! How far "fairness" has come in 81 years!

The Darwinists in 1925 at the Scopes Trial really *were* sincerely interested in Creation and Evolution being taught side by side in school.

Weren't they?

GOD'S NOT DEAD (DARWINISM'S KILLING SCIENCE)

Say you live in a house and everyday a new crack appears in the walls, the floor, the ceiling. You notice the windows are beginning to sink to ground level, and the plumbing and wiring is snapping from the tension of the sagging structure. What do you conclude? The house is sound? Everything's OK?

The house of science is cracked, falling to pieces. It's got the wrong foundation. It was originally built on the foundation of Biblical revelation, especially Creation. Read about the founders of modern science like Francis Bacon.

But in the late 1850s the house of science moved onto a new foundation: Darwin.

But there is not a single solitary advance in science that comes from Darwin's Theory. The advances in genetics have nothing to do with belief in primordial soup. The advances in chemistry don't need the theory of frogs becoming dogs. The advances in space technology do not rest upon the big bang. The advances in medicine do not rely upon the dogma of apes coming out of the trees.

What are we saying? *That Darwin has contributed nothing to the advancement science, only the advancement of godless Naturalism*. Naturalism exists because people do not want to deal with a God who can tell them their behavior is wrong. Simple. So they live in a ramshackle house built on Darwinian evolution instead.

Ask yourself, and be brutally honest, **what would change in the impact or the advancement of science if Darwin were no longer taught as science?**

IRREDUCIBLE COMPLEXITY: LEARN IT, LOVE IT

Our Creator did not intend for us to wonder for a moment whether or not he made everything. He left so many clues it would make your head spin.

For example, take Michael Behe's recent brilliant book about of the principle of "**irreducible complexity**." Behe is a microbiologist who's done painstaking examination of the tiny molecular machines that make up life at the cellular level. The complex systems upon which life is built could not function in simpler configurations. They can't be reduced. They're *irreducible*. Hold that thought.

Darwin did not know about the structure of cells. He hoped cells would be absurdly simple things like jello that could just have happened through a chemical accident. But! It turns out that cells are no less complicated than the human body they form. Fantastically detailed molecular brains and factories and security systems and energy transfer stations exist in the vast reaches of inner space seen only by microscopes!

And these complex systems are **irreducibly complex**. Take away any one of their subcomponent parts and they don't function at all, they're useless.

Cells, the building blocks of life, are unmistakably, irrefutably *designed*! They could not possibly have formed by accident because they are incredibly complicated and every tiny part is necessary for the whole to function! Life, even at it's simplest, irreducible micro level, is a miracle.

If Darwin knew then what we know now, he'd have to be an idiot to even propose evolution. Makes you wonder about people who still believe it, knowing what we now know about cells. Doesn't it?

CAN YOU THINK OF JUST ONE THING?

Colin Patterson, evolutionist and Senior Principal Scientific Officer in the Paleontology Department of British Museum of Natural History in London, from 1962-1993 was asked to tour America to speak about evolution. He says, "One morning I woke up and something had happened in the night, and it struck me that I had been working on this stuff for twenty years and there was not one thing I knew about it. That's quite a shock to learn that one can be misled so long. Either there was something wrong with me, or there was something wrong with evolution theory."

To nationwide audiences of people deeply involved in science, he would ask a simple question. He want someone to tell him *"anything you know about evolution, any one thing, any one thing that you KNOW is TRUE?"*

Patterson later recalled, *"I tried that question on the geology staff of the Field Museum of Natural History and the only answer I got was silence. I tried it on the members of the Evolutionary Morphology Seminar in the University of Chicago, a very prestigious body of evolutionists, and all I got there was silence for a long time."*

Darwin's theory has had all the advertising muscle of the entire scientific community behind it for well over a hundred years. And still there's not *one thing* about it that a group of Darwinists can claim is inarguably true! This is stunning. The evolutionists laugh at anyone who remotely questions the veracity of their cockeyed idea and yet among themselves they cannot offer up a single solitary proof that veracity.

"There is something remarkable about science. It gets such wholesale returns of conjectures out of such a trifling investment of facts."
– Mark Twain.

EMBARRASSMENT OF THE EXISTENCE OF SPECIES

Think for a moment about dogs and cats. They belong to different biological groups. There's a distinct line they cannot cross. This dividing line is called species.

Puppies come from a mommy dog and a daddy dog, not a mommy dog and daddy cat. That's the distinction of species: the ability to reproduce. I'm sorry to sound so sophomoric, but evolutionists simply skip over this easily understood scientific maxim as if it weren't even there. They think variation *within* a species proves that species can evolve into *other* species. What they believe is more sophomoric than saying mommies and daddies. They believe something no one else does: that because we have puppies, evolution is now a proven fact.

But think, why even HAVE species if life *evolved*? Why would we not have a continuous blur of organisms that all look sort of the same and can all interbreed? **Why are there distinct species? And why is the distinction especially evident in the fossil record** which should be the "blurriest" of all? This question is not new. Darwin himself wondered, "*Why, if species have descended from other species by insensibly fine gradations, do we not everywhere see innumerable transitional forms? Why is not all nature in confusion instead of the species being, as we see them, well defined?*"

Though we already have rooms filled with countless fossils that we don't have time to look at because we are constantly digging up more — maybe that's still not enough to see the hoped for "innumerable" links? I bet we'll find the missing link if we keep trying! Maybe we just need to persevere. Maybe its because we don't have enough evidence yet!

Or maybe it's because evolution is stupid nonsense.

NOTHING SO ABSURD AS NUMERICALLY ABSURD

Super scientists Fred Hoyle, Chandra Wickramasinghe, and Francis Crick are high profile atheistic naturalists, possessing multiple degrees both scholarly and honorary and have won the highest awards given to scientists of their respective fields.

One of them, now dead, was an astrophysicist (read "rocket scientist"), one is an astronomy and applied mathematics professor, and one discovered the DNA helix. Their credentials are unassailable. They cannot be laughed off as Bible-thumping hayseed evangelical pea-brains.

None of them buy Darwin. That doesn't mean they buy creation, but they know enough science to see that evolution is silly. You'll see what they DO believe below!

Hoyle calculated the probability of producing the basic enzymes needed to even begin to build something remotely resembling organic at 1 chance in 1 followed by forty-thousand zeroes!

Crick's description of the odds of evolution happening, "So utterly minuscule as to make it absurd."

A well known atheist brainiac named Stephen Hawking said, "The odds against a universe like ours emerging out of something like the big bang are enormous... Clearly there are religious implications whenever you start to discuss origins of the universe."

And yet, though evolution on our planet is mathematically impossible, these scientists still won't consider God created life. Hoyle, Wickramasinghe, and Crick believe aliens sent life to planet earth in space ships. *Seriously*. Not joking!

APPENDIX 4: DIGGING DEEPER

THE FITTEST ARE THE ONES WHO SURVIVE

As far as a logical argument, Darwin's "Survival of the Fittest" is fatally flawed. It's a "**tautology**," a self-fulfilling prophesy, a statement that relies upon itself to be true.

If you say that the mechanism of evolution is the survival of the fittest and the fittest are defined as the survivors, then of course the fittest survive since that's how they're DEFINED! Hello.

Darwin's theory is the survivingness of the survivors! Since "fittest" simply means whatever an organism has that allows it to survive. Not much of a breakthrough in Western thought, is it? This theory says survivors survive and upon that statement, entire branches of science stand or fall.

More slipshod reasoning is found in the Darwinists' use of fossils to determine the age of rock layers. It's lovely to find a Brachiopod if you're a geologist, because you know you're in the Cambrian epoch of time. And how do you know Brachiopods are from the Cambrian? They're the same age as Cambrian rocks. And how do we know they're the same age? Because we dig them up together. So we're saying that the fossils tell us the age of the rocks while the rocks tell us the age of the fossils. Riiiiiiight. The fossils are old because they are in old rocks that we know are old because they contain old fossils. It's all neat and clean.

What's going on here is called "**circular reasoning**," and what passes for science these days would never make it in debate class. To let the "fittest" be the ones who "survive" and to let the rocks and fossils assign ages to each other is the very definition of circular reasoning and draws a penalty flag from any critical thinker.

Evolutionary proofs are tautological nightmares.

ALL SCIENTISTS BUY EVOLUTION, DON'T THEY?

"Scientists who go about teaching evolution as a fact of life are great con men," says Theodore Tahmisian, a nuclear physicist. *"In explaining evolution, we do not have one iota of fact."*

"Evolution is still a highly speculative hypothesis entirely without direct factual support," says Evolutionist Michael Denton, *"and far from that self-evident axiom some of its more aggressive advocates would have us believe."*

"Overwhelmingly strong proofs of intelligent and benevolent design lie around us," says Lord Kelvin, an icon of scientific accomplishments. *"The atheistic idea is so nonsensical I cannot put it into words."*

"The attempt to explain all living forms in terms of evolution is premature," says G.A. Kerkut, physiologist and evolutionist, *"and not satisfactorily supported by present day evidence."*

"Evolution is a fairytale for grownups," says Dr. Louis Bounoure, Director of Research at the National Center for Scientific Research in France.

All scientists believe in the law of gravity, the laws of thermodynamics, Newton's 3 laws of motion and Kepler's laws of planetary motion. Sounds like not all scientists buy the "law" of evolution. They don't even buy the theory of it.

What might that tell us? For one thing, it says that if you don't believe in evolution, if you just can't bring yourself to think that life came from rocks in hot water no matter how many textbooks and TV shows and "smart" people tell you it's true... then you have good company with a lot of scientific minds. Truth doesn't need textbooks and TV shows and "smart" people telling us it's true. There are truths that are so true they are "self evident." Like Creation.

THE ODD SUBJECT OF ODDS

The odds that you will be born as a twin are 90 to 1.

Your odds of becoming a pro athlete? 22,000 to 1.

The odds of dating a supermodel? 88,000 to 1.

Your odds of getting killed by lighting? 2 million to 1.

The odds of a meteor from outer space hitting your house are 182,138,880,000,000 to 1. If you're scared that someday a space rock will kill you, get over it.

We already told you that evolutionary astrophysicist Sir Fred Hoyle calculated the probability of producing the building blocks of life, not even life itself, by mere chance at 1 followed by 40 thousand zeroes to 1 and quipped, *"The chances...are comparable to a tornado ripping through a junkyard and producing a Boeing 747."*

What Sir Fred was saying in his dry humorous way was that there IS a chance that life could evolve, but the *odds* are that it didn't. In fact, a tornado *could* rip through a junkyard and actually produce a Boeing 747. But I wouldn't hold my breath waiting for it to happen. The chances of that happening are the same as the chances of evolution being true.

So don't bet on evolution. The odds are prohibitive.

By the way, your odds of eventually dying are 100%. The evolutionists are going to find out that the Creator has a lot to do with what happens to them next.

Again, 100% certainty.

THE ODD SUBJECT OF ODD PEOPLE

The dictionary defines *"odd"* as: different from what is usual.

Stephen Wiltshire is odd. He's autistic and considered brilliantly handicapped, an idiot savant.

Stephen's first words at age 5: "pencil" and "paper." His nickname is the human camera. He can see something once and draw it precisely without ever looking at it again. At age 11, one helicopter flight over the city of London gave him all the information he needed to draw the city down to the exact number of windows.

He was taken on a helicopter ride over another city he'd never seen before: Rome, one of the most ancient and complex cityscapes in the world. After his 45 minute viewing, Stephan was given a 17 foot sheet of drawing paper and only 3 days to draw a panorama of the city. With no preliminary sketching, he inked in a near perfect representation of the city down to the hundreds of small houses crowding the winding streets, numbers and height of columns, the broken down ruins and the appearance of the river, including St. Peter's, the Pantheon and the Colosseum.

Darwin can give us no clue to explain this man. The brain alone is a mystery to Darwinists, let alone the capacity of the fat cells in our heads to store such detail in our memory and then be able to recall it perfectly as Stephen Wiltshire can.

The truth is that our Creator fashioned this marvelous person in his mother's womb, not in primordial seas. And God has fashioned each of us with unique gifts and potential because he never runs out of creativity.

I like that!

APPENDIX 4: DIGGING DEEPER

SPEAKING OF THE ANIMAL CHANNEL

Bees communicate by dancing and they can fly even though their wings should be inadequate for their body weight.

Spiders with tiny mouths turn their larger prey into drinkable smoothies. Whales, the largest of animals eat plankton, the smallest.

Fragile butterflies migrate thousands of miles. Venus fly traps are plants that catch and eat animals by subterfuge and disguise.

Coral reefs are animals that build entire islands using only their own dead bodies. Ants have well defined job descriptions and society castes. Cats can safely fall 50 feet and land on their feet after acrobatically adjusting in midair. Bombardier beetles explode 212 degree substances from their business end. Bears can sleep for half the year at a time. Flies can land on the ceiling.

Frogs swallow their food by pushing it down their throats with their own eyes. Some mollusks eat their arms and then re-grow them. Horses and cows sleep standing up. Sheep have two stomachs. Camels can go without water longer than you'd think possible.

Penguins are birds that can't fly through the air but can fly through water faster than fish, while wearing a tuxedo. Crocodiles let delicious birds safely eat the food between their teeth.

I could go on. The clever naturalists say all this brain-exploding creativity is just a freak accident that came from rocks in hot water over eons of time. The naturalists are really not so good at observing and forming right conclusions.

PEOPLE ARE NOT SMARTER THAN GOD

I used to think I was smarter than God.

My elementary school science class did an experiment with fruit flies. We ended up with fruit flies that looked different from their parents. Being smarter than God, I didn't stop to observe that I myself looked different from my parents and in fact every single creature on the face of the earth looks different from their parents.

The teacher could have pointed out that God gifted our genetic makeup with His usual extravagance and the myriad possible combinations of those genes easily explain the wonderful variety within species. That would have been actual science.

But the teacher also thought he was smarter than God and instead told us that the experiment proved we evolved from primordial soup. Hooray for us, idiots all.

Same thing with dinosaurs. Kids love dinosaurs. I loved them too. Monsters! Kids love monsters. Especially when they're made of plastic hand painted in Taiwan. Adults who were smarter than God told us dinosaurs evolved. We bought it. No need for an explanation, or any chain of reasoning. Just believe in evolution because of dinosaurs and fruit flies.

Dinosaurs and fruit flies! Get it?!! No? You say it's silly?

You don't see how dinosaurs and fruit flies prove evolution? Try again — there's a dinosaur skeleton — *now* do you see evolution? Just look. *Dinosaur*. DINOsaur. Evolution. Dinosaur. It's so *obvious*.

You have to first be smarter than God and then you'll get it.

EOHIPPUS: THE DOG THAT BECAME A HORSE

You've seen a chart that appears in nearly every single textbook about evolution that shows a dog-like horse morphing over time into our "modern horse," tall and strong.

Marvelous. Diggin' up bones, drawin' up charts.

So now evolution is proved beyond doubt. We have an actual *chart* somebody drew to prove dog-horses became horse-horses! The mighty pen! If you can draw it, it must be true!

But if the different horses evolved in perfect sequence through time, how come we never see them in their supposed sequence in the rock layers?

How come the horses in the chart go from 18 pairs of ribs to 15 to 19 then back to 18?

How come some of the "horses" are in fact hippopotamuses?

Maybe, just maybe, God created a bunch of different kinds of horses and most are now extinct. *That* would make sense.

You see, Creation explains the fossils, evolution can't. While it's *clever* to draw a little chart to claim horse fossils prove evolution, it's not honest.

"Classic cases such as the evolution of the horse have had to be modified or discarded as the result of more detailed information." David Raup, Curator of the Field Museum of Natural History in Chicago.

Next time you see the horse family chart offered as proof of evolution, laugh quietly.

A hoarse laugh, if you will...

THAT'S WHY WE CALL THEM MISSING LINKS

You've heard of the missing link?

The unintended admission by the naturalists who talk about missing links is that the species we dig up in the fossil record are not connected in any way, or else we'd not be looking for links. If they are not connected, how come we have these beautiful "trees" that show single cell animals at the roots, branching into fish and then amphibians and then reptiles and the birds and then mammals and finally people?

Mark Golblatt in USA TODAY, July 2000, wrote, *"The theory of evolution is supported by mountains of scientific evidence."* The only problem is, these are *invisible* mountains that no one can seem to find! Not even small foothills of evidence are around.

"Very few intermediates between groups are known from the fossil record," offers Robert Carroll, evolutionary biologist and paleontologist. "Very few" actually means "none!"

"It is not even possible to make a caricature of evolution out of [the fossil record]," points out Professor N. Herbert Nilsson.

"The family trees which adorn our textbooks are based on inference, not on evidence," laughs Steven Jay Gould, rabid evolutionist from Harvard U. He also admits the "extreme rarity" of transitional fossils is one of the "trade secrets of paleontology."

"Nine-tenths of the talk of evolutionists is sheer nonsense, wholly unsupported by facts. In all this great museum, there is not a particle of evidence for the transmutation of species," declares Dr. Etheridge, paleontologist of the British Museum.

"The whole reason that we have missing links is because they're missing," quips Ken Ham, science teacher and lecturer.

TAKE A SHOVEL FOR A TEST DRIVE

Listening to evolutionists you'd think that anywhere we dig in the earth, we first find fossils of large mammals, then deeper down we come across dinosaurs, then deeper bony fishes, then shell fish, and finally the famous little trilobites.

But you may be surprised to learn that that's not what we actually DO find. We find all the fossils jumbled together and mixed up in the most perplexing way.

Animals believed to have been extinct millions of years are found with those supposed to have recently evolved. Some animals believed extinct appear AFTER their supposed extinction (or found to be still alive in the case of the Coelacanth) and some appear long before they were supposed to have evolved.

It's a mess! The fossil record *seems* to show that all the various species originally coexisted on the planet and since then, some have become extinct!

Often it appears that some of the fossils were sorted by flood waters and they settled in sediment-loaded waters by size and density. Like in a flood. Hmm. Was there a flood? Anyone remember a world wide flood?

For the evolutionist facing these unruly fossils in the rock record, they solve this problem by ignoring it and by throwing out the fossils that don't agree with their theory.

Because evolution is the FACT. The evidence is unnecessary. Anyway, it's the fossils that are confused.

Everyone knows fossils make mistakes — after all, some are only human.

EXACTLY HOW MANY BYTES IN A TRILOBITE?

Trilobites are what geologists call an *index fossil*. An index fossil is only found in rocks laid down in a certain time period, so if you see that fossil, you can tell how old the rock layer is that you are digging in. Pretty convenient.

One day geologists found trilobite fossils and said, "Wow. This rock is really old. See? TRILOBITE FOSSILS. Trilobites are really old, so the rocks must be really old too."

If you're scratching your head why trilobites have been certified as some of the oldest critters on earth, so am I. How do we know that? Do the trilobites have tags tied around their shells with their birthdays? Nope.

Evolutionist say they are old, because they need to have small old things on the bottom of the ocean long ago for bigger things to have evolved from. No one can deny that trilobites are small things on the bottom of the ocean.

So the trilobites are old because old small things must have been on the ancient sea floors, and since trilobites live on the ocean floor and they're small, hey presto! They're old. Don't ask how anyone could possibly know that.

And certainly don't ask the live trilobites we see today on the ocean floor how finding them in a rock layer means we are in 450 million year old rocks. Shhhh. They're extinct. They're an index fossil that's only found in old rocks. So they *can't* be alive today. That would make us doubt evolution.

The Coelacanth is used like the trilobite as an index fossil: if you found a coelacanth, you knew you were in rock at least 65 million years old. Until 1938, when someone pulled a live one out of the Indian Ocean.

THE CAMBRIAN EXPLOSION BLOWS UP DARWIN

Did you know that the fossil record has a dirty little secret? Fossils are found in rocks to a certain depth and then...nothing.

Not even a loose flagella from a balding amoeba.

There are fossils preserved in the rock layers until you reach a particular depth and below that: nada.

It's as if life began all at once (!) instead of spread out over great epochs of time.

It's called the Cambrian Explosion and for good reason. This amazing bedding plane in the rock is the null point between all of life and no life at all. Below this point, there is no indication of life in the rocks. Above this point, every known major plant or animal group is represented.

There was nothing and then all of a sudden, blam! Every living creature shows up. If there's one true thing about the fossil record, it's that we have rocks with no record of life underlying rocks with dead animals buried in them. Beyond that, every other observation is a wild-eyed guess.

So how to explain the sudden appearance of fossils in the rock record? Are we missing 4 billion years worth of fossils before the Cambrian Explosion?

Or is it like the Bible says – God made everything in one creative wave of his hand — and the fossil record proves it?

Never wonder we call modern naturalist science a religion: they have faith in fossils they cannot see.

SACRED TEXTBOOK DRAWINGS PROVE EVOLUTION

It's your solemn duty to bow anytime you see a chart with small sea creatures on the bottom which evolve into fish, then frogs, then birds, then dinosaurs, then mastodons, and finally apes leaving the trees and becoming your grandfather.

Marvel at the imagination of the Darwinists. The holy chart doesn't need data to support it's claims. Look upon it reverently and remove your curiosity.

Nowhere in all of God's creation is there a fossil bed that matches this chart or any part of it.

Nowhere in the rock record is there the smallest glimpse of what the chart plainly asserts.

Nowhere in all of nature's vast treasure house is there the least indication that what the chart tries to tell you is worth a wooden nickel. But that doesn't mean the chart isn't true. It just means we don't have enough data. YET.

So until we come up with the corroboration for this chart, of course we'll use it. We'll teach it to small children and to college students. We'll write endless theses on whole forests of paper about the wonderful chart. We'll make documentaries that use the chart to say all kinds of things about which we have no clue. We'll color the chart and update it and make sure that every Geologist worth a paycheck knows it by heart.

And we will defend the chart against all attacks of logic, common sense and calls for supporting data.

Truth is for sissies.

This is THE CHART.

THE BOMBARDIER BEETLE RED FLAG

Among the scads of amazing creatures on this planet is an insect called the Bombardier ("Bohm-bah-deer") Beetle.

If you are a hungry predator with an idea to eat this little guy, you're in for a rude awakening.

No sooner do you open your mouth to nab him for a quick snack than he explodes 212 degrees worth of boiling hot chemicals in your face. If that's not an appetite suppressant, I don't know what is!

How did he evolve? Easy. All you need to build him can be found on any standard-issue bug. Simply retrofit the base model beetle with a chambered gland that manufactures the specific chemicals hydroquinone and hydrogen peroxide. No problem, most glands do that, I guess, right? Then evolve the gland on his rear end so he doesn't explode the chemicals in his own face. Then evolve some way of inhibiting the reaction until the Bombardier Beetle is about to be eaten and at that point evolve him an anti-inhibitor to inhibit the inhibitor so the explosion can take place.

And since the predator doesn't always approach from the rear, evolve the whole apparatus on a turret that can be aimed in any direction and evolve each bug a set of detailed instructions how to use the rear-end-explosion-maker-hot-bomb-defense thing.

To the evolutionist, this is no problem. *Because the Bombardier Beetle exists, he must have evolved.*

No design or Designer required. Isn't it astoundingly remarkable what space rays hitting wet rocks can do?

APPENDIX 4: DIGGING DEEPER

THE KILLER ARGUMENT THAT ENDS ALL DEBATE

I used to think that if I could just find that killer argument I could shake and wake a friend to see that Darwin is stupidity.

I spent years reading and writing down the brilliant ideas I read that blew Darwin out of the water. There are hundreds of them and every one of them totally refutes Darwinian evolution to any open, thinking, truth-seeking person.

And what did I find? The killer argument just doesn't exist. Not because there aren't fabulous refutations of Darwin, there are, and they ARE killer, but because only God's Spirit can awaken a deceived heart.

I should know, I had one of those deceived hearts myself.

There are killer arguments from design, and killer arguments from logic. Killer arguments from math and probability and killer arguments from astronomy and physics. There are killer arguments from genetics and microbiology. There are killer arguments from history and killer arguments by simply quoting distinguished scientists. There are killer arguments from ancient proverbs and killer arguments from common sense. Don't even get me started on the reams of killer arguments from God's wisdom revealed in the Bible.

If someone has been duped, it's only by a miracle that they become UN-duped.

As it says in 2Corinthians 3:16, **"To this day, a veil remains over their hearts, but when anyone turns to Christ, the veil is taken away."** So here is the only fool proof approach:

Pray for your friends. Love them and never quit praying.

PLEASE PASS THE CLUELESSNESS

In talking with one of my dearest friends, I found he didn't understand the claims of Darwin, but thought he did.

My friend had grown comfortable with a fuzzy idea of evolution and couldn't believe anyone would ever question it. In his mind, evolution and the Bible worked together — one was science and one was religion. There was no conflict.

"The Bible explains why and Darwin explains how," he mindlessly regurgitated from somewhere. He took Scripture seriously but thought the Creation account was a sort of metaphor.

He had no problem questioning Genesis, but didn't believe we should ever question Darwin. He mumbled something about Darwin being science and science is honest, objective and unbiased.

But science is always conducted by scientists. *And scientists are people.* People can be closed-minded, agenda-driven, under-informed, sinful, prideful, gutless clueless and wrong.

He couldn't see that Darwin is a dogma with articles of blind faith the same as any other religion. He thought science supports Darwin, but science is observable, demonstrable and falsifiable. Darwin is none of those.

He thinks the Bible has to prove it's claims beyond a shadow of doubt, but would never dream of holding Darwin to that standard.

My friend is a good person with confused ideas of God because the Trojan Horse of Darwin has planted a destructive virus in his mind.

EVOLUTIONARY TIME CAN'T DO MATH

I have a nifty cardboard chart of evolutionary time. I wish you could see it.

Around the outside of the circle are fictional epochs of prehistory. As you spin a dial to point to these epochs, the chart neatly tells you which plants and animals evolved during that period of time. For example, if you turn the dial to 500 million years ago, you line up with the Ordovician Period and you can see drawings of a cephalopod, a brachiopod, and a sea lily.

Wonderful. nothing like a great chart to indoctrinate people in a godless naturalistic religion!

As well as being fraudulent, the wheel has the added bonus of having no corroborating scientific data, a deficiency more than offset by the cleverness of the gizmo itself and the cute little drawings.

And if you read the blurb in the center of this dazzling devise, you'll find that it's only a snapshot of 10% of earth's history, that the earth is really FOUR AND A HALF billion years old.

Don't ask how they know that, you couldn't possibly understand. Just believe it. And the beginning of life reads out at 600 million years.

But wait. To get 100% of the earth's age, you multiply 600 million (10%) by 10 and you get SIX billion years.

The chart isn't aware that it says the earth is both 4.5 billion AND 6 billion years old. I guess 1.5 billion is just a rounding error.

Science is never wrong. *By much.*

EVOLUTIONISTS BELIEVE WITHOUT SEEING

Darwin was cheerful about the glaring flaw in his theory.

And here's the flaw: he knew that what he proposed had never been observed, but he had hopes that the evidence would show up. His main hope was in the fossil record.

So he was candid about the true state of things. *"Not one change of species into another is on record."* He wasn't worried. The proof would be found. It was only a matter of time and finding a few more fossils.

But if the truth be told, the fossil record has been a bitter disappointment for the Darwinists. Now and then an isolated fossil will be found and proclaimed as proof for the evolution of species. But these few discoveries never provide the conclusive proof needed for such a sweeping theory as Darwin proposed. They're just a few examples open to wide, and I mean *reeeeealy* wide, interpretation.

Archaeopteryx was such a fossil – hey! A reptile with feathers! But there are no fossils that link to it. What's it a transition from or to? It is now considered just a bird and not a reptile.

The confidently hoped for "mountains" of irrefutable missing links are reduced to a couple of "maybe if you close one eye" links.

I remind you of this quote: *"It is not even possible to make a caricature of evolution out of [the fossil record],"* points out Professor N. Herbert Nilsson.

And this one: *"The family trees which adorn our textbooks are based on inference, not on evidence,"* laughs Steven Jay Gould, rabid evolutionist from Harvard U.

EVEN IF IT WERE TRUE, WHO WOULD WANT IT?

Psalm 100 says, **"Know that the Lord is God, He has made us and we are His people."**

The evolutionist says, "Know that you are a freak accident, here for no reason, to survive as best you can until the time of your meaningless death when you return to the nothingness from which you came."

Forget for a moment which statement really IS true, which statement would you *wish* to be true?

And then think. *Could* you conceivably be able to conceive the first statement if in fact the second statement WAS true?

Could you muster up the Technicolor dream of such a wonderful thing as a Creator and His love for you if you're nothing more than a freak accident of hydrogen gas? Can a random collection of molecules even invent a wonderful Creator God who loves us and wants us to live with Him forever? I doubt it.

If God is only wishful thinking, then give me wishful thinking any day. Paraphrasing Puddleglum talking to the Green Witch of the underworld in C.S. Lewis's book *The Silver Chair*, "Maybe the sun and the rain and the sky and Aslan are all only a dream, but if so, it seems that our dream world beats your real world hollow."

"When I think of the wisdom and scope of God's plan, I fall to my knees and pray to the Father, the Creator of everything in heaven and on earth." Ephesians 3:14-15. Do you just *want* to be alone in the universe, adrift in the gasses of space?

Or know your Creator who made you and loves you?

WELL MY WORD, WHAT A CO-INKYDINK!

How lucky are we? *"It turns out that the constants of nature have exactly the values that allow stars and planets."* Physicist Edward Kolb of Fermi National Accelerator Laboratory.

We live in an impossibly improbable universe: *"Every one of nature's forces must have just the right strength. If electrical forces were stronger, no element heavier than hydrogen could form, if electrical repulsion were too weak, the sun would explode."* Physicist Richard Morris, on the lucky accident of the forces of nature.

It's a good thing for us that we don't have a moon heading straight for us: *"If gravity were too small, it would not turn the moon out of a rectilinear course: if too great, it would draw the moon towards the Earth."* Sir Isaac Newton, on the lucky accident of moons.

Everywhere we look, God's handiwork is dazzlingly awesome: *"Everything in the universe is revolving as if in a never ending waltz. There is no natural reason why the outward force produced by revolving should so perfectly counterbalance gravitation."* Fred Heeren on the lucky accident of orbits.

And if naturalists really want to wrap their minds around the cosmos, the very first thing they need to do is stop being naturalists!

"We can't understand the universe in any clear way without the supernatural," quips astronomer Allan Sandage.

Understatement of the decade.

Class dismissed.

NOT GORILLAS, YOU IDIOT, APE-LIKE CREATURES!

I am amused by the Darwinist's exasperation when their theory is summarized as "Your great grandfather was a monkey."

"No, no, no!" they protest, "Not a monkey, an *ape-like* creature." "Aren't monkeys ape-like?" "Yes, but your grandfather did not descend from monkeys." "Who did my grandfather descend from?" "We don't know exactly." "So not ape-like *monkeys*, but ape-like something elses?" "Yes, that's it!" Oh gag me.

Anthropologists, using billions in tax dollars taken from taxpayer pockets, have tried mightily to prove Darwin and have come up with absolutely nothing. Zero. Drawn a blank. Gotten a goose egg. Failed. Fumbled. Whiffed. Missed. I'm not saying these things, THEY are! Listen:

"Virtually all our theories about human origins were relatively unconstrained by data. The theories are fossil FREE. In some cases, even fossil PROOF," admits Richard Pilbeam, anthropologist, revealing that when lowly scientific facts stand in the way of an Anthropologist's guesses, the guesses win every time.

Before Darwinists blast us for daring to ask for support for their theory, they'd check the volume of evidence available. Like the supposedly solid evolutionary tree of man: *"All the skeletal evidence of early man would barely cover a billiard table...preconceptions have led evidence by the nose,"* acknowledges John Reader, anthropologist.

I thought there were "mountains" of evidence. I can't imagine a mountain on a pool table...

TIME FOR THE RADIOACTIVE ISOTOPE DATING GAME

It's a fact. Electrons sometimes spin out of control as they navigate their way around the nucleus of an atom. When you're talking *radioactive* isotopes, this happens a lot. The radioactive elements are by nature less stable.

Now let the games begin! Using radioactive decay measurements, Naturalists can guess precisely how old a fossil is.

It seems to not matter to them that every time we can cross check their findings, their findings are way off. But that's OK right? Since evolution has to be true, the fossils have to be old, and since they have to be old, the radioactive isotopes prove it. Even if they don't.

Anyway, when we use isotope dating methods, *there are a few wild assumptions we need to make*. To use the radioactive isotope method, we must assume that the decay rate for a radioactive isotope is constant *(although there's no way to know that)*, and assume that the sample began with a certain amount of the isotope *(although there's no way to know that)*, and then assume that no heat or pressure has been applied to the sample *(although you know it has or it wouldn't be a rock)*, and assume that the Law of Uniformity, look it up, is in operation *(even though everything we can verify says it's not)*.

Oh yes, you also must assume that your sample has not been anywhere near other radioactive elements *(too late! It's been near itself!)* and assume that it absolutely HAS to be VERY OLD. Most importantly for evolutionists, disregard any conflicting results as "anomalous."

Then you can't miss! Isn't science wonderful! It takes much more faith than believing the Bible ever could!

"ACCIDENTAL" PURPOSEFUL DESIGN?! WHAT?!

We said previously that DNA, the fantastic molecular supercomputer in each cell, is a genetic *language* with subjects, verbs, predicates, syntax and punctuation.

Why is the existence of DNA fatal to Darwin's theory?

Julian Huxley famously claimed that if you had enough monkeys, enough typewriters and enough time, they could come up with a library full of brilliant books by randomly pecking away.

Sounds good. But it's absolute nonsense. *"Huxley was hopelessly wrong in stating that 6 monkeys allowed enormous time would be able to randomly type all the books in the British Museum,"* stated Cyberneticist David Foster, *"when in fact they could only type half a line of one book if they typed for the duration of the Universe."*

Foster was being too kind. DNA is a *language*, and there's no such thing as a randomly generated language. It's an oxymoron.

How many times would you have to knock your computer off the desk before it output a Doctoral Thesis on orbital velocity in Latin?!

Why would any rational person believe that DNA was randomly generated from an explosion of gas? Why would someone with a brain believe that accidental language is possible? Only because they wanted so badly to believe it that they were willing to sell their soul and shut down their God-given scientific skepticism.

Language in DNA is a dead giveaway that there is a highly intelligent Creator of that language.

CHRISTIAN DARWINIST: A WALKING CONTRADICTION

A Christian who believes in evolution is confused. This mangled philosophy is called Theistic Evolution. There are people who will tell you that the Bible and Darwin are in harmony. They will plead with you not to take the Bible literally. They have a problem with the Bible:

- Genesis 1:1 **"In the beginning God created the heavens and the earth."**
- Exodus 20:11 "**For in six days the LORD made the heavens and the earth, the sea, and all that is in them."**
- 2Peter 3:5 **"They deliberately forget that long ago by God's word the heavens existed and the earth was formed."**
- Acts 17:24 & 26 "**The God who made the world and everything in it is the Lord of heaven and earth. From one man he made every nation."**
- Romans 1:18-22 **"People know God's truth instinctively. From creation, they have seen God's handiwork, but they wouldn't worship Him. Claiming to be wise, they became fools."**
- Colossians 1:15-17 **"For by him all things were created: things in heaven and on earth, visible and invisible; all things were created by him and for him. In him all things hold together."**
- Job 38:36 **"Who endowed the heart with wisdom or gave understanding to the mind?"**
- John 1:3 and 10 **"Through him all things were made; without him nothing was made that has been made. Though the world was made through him, the world did not recognize him."**
- Revelation 4:11 **"You are worthy, our Lord and God, to receive glory and honor and power, for you created all things, and by your will they were created."**

How can a Theistic Evolutionist be a Christian while ignoring plain-as-rain Biblical doctrine? They can't.

APPENDIX 4: DIGGING DEEPER

SOME OF MY FAVORITE BOOKS ON THIS STUFF

Always begin with the Bible, the measuring stick of truth. The Bible is the benchmark from which we can judge all ideas. It's important to have God's word firmly in our minds so we can recognize truth when we meet it in the world. People who must detect counterfeit money study the real currency: then when anything else comes along, it's evident immediately. Same thing with truth: Study the Bible so you recognize ideas that cannot be true.

Remember that we Christians love science, all the more because we love the God who invented a universe governed by his laws of science! We want to find his fingerprints.

Here are some books you'll enjoy. I don't always agree with these guys, nor they with me! It's OK to read people who have different ideas, right?

- UNLOCKING THE MYSTERIES OF CREATION, D. Petersen
- GODLESS: Ann Coulter (she's great on this subject!)
- SHOW ME GOD, Fred Heeren (a science journalist)
- GUIDE TO HOW IT ALL BEGAN, B. Bickel and S. Jantz
- THE GENESIS SOLUTION, Ken Ham and Paul Taylor
- THE COLLAPSE OF EVOLUTION, Scott Huse
- DARWIN'S LEAP OF FAITH: J. Ankerberg and J. Weldon
- SHATTERING THE MYTHS OF DARWINISM, Richard Milton
- THE CREATOR AND THE COSMOS, Hugh Ross
- THE GENESIS FLOOD, J. Whitcomb and Henry Morris
- Phillip E. Johnson has tons of thoughtful books like: DARWIN ON TRIAL, REASON IN THE BALANCE, and THE WEDGE OF TRUTH

QUESTIONS TO POLITELY ASK EVOLUTIONISTS

- Do we know the origin of matter and energy?
- Is the speed of light a constant? How can we be sure? When did we start keeping records of the speed of light and how accurate are those measurements?
- If the speed of light is not constant, how old is the universe? Can we know?
- How many dropped computers would it take to write a doctoral thesis on bacteria in perfect French? How probable is it that sun rays disrupting DNA sequences could produce a viable life form?
- Is the geologic column present in it's entirety anywhere in the world? Who decided that it exists?
- Can fossilization happen quickly? Since that answer is yes, why is fossilization assumed to always indicate great age?
- What caused the Cambrian Explosion?
- What is the definition of species? In the last 500 years, how many new species can you name?
- Why did Darwin's original theory need to be revised? When will a revision occur to the new theory?
- Would you mind telling the complete story of evolution from the Big Bang to Neanderthal man? Which points of your story are supported by reproducible, falsifiable, demonstrable science?
- What is an axiom? Since an axiom cannot be proven, and the Principle of Uniformity is an axiom, why is scientific faith different from faith in God? Is an axiom a matter of science or faith?
- Correctly name the axioms of mathematics.
- What is an assumption? How many assumptions do you have to make for evolution to be true?

APPENDIX 5: MORE AT WEBSITE

Other terrific books in the series at **www.alanroot.com**

Designamentals
You're not an accident, you're not an animal, you're the masterpiece of a Master Mind

Cryptoleganomics
One world religion has a Savior... the secret of why Jesus is the only way

Salvificology
Salvation: it's so much more than just going to heaven when we die

Disciplification
The way a master teaches is not through rules but through relationship

Bibliotectonics
Exploring the most loved most controversial most earth-shaking book in the world

Orthodoctrines
The rock-bottom non-negotiable teachings that all Christians have always believed

Alienpilgrimometry
What's left of our faith after it slams into the frenetic humdrum of our daily lives

www.alanroot.com Search for Alan Root on iTunes for rockin' kid tunes! *Ask My Pastor, He Knows That Stuff*, Alan's new book for all ages about Biblical theology, is now available on Amazon.

Don't ever forget Jesus loves you. He's our Maker and our Master. He became one of us, died on the cross, rose from the dead, never to die again. And he invites us, each one, to repent and follow him, by the power of the Holy Spirit, into the purpose-packed and satisfying lives we were always meant to live, here in this world and forever in the ages to come. In Him we live and move and have our being. In Him we find mercy, love, and grace for each day!

Made in the USA
Monee, IL
04 November 2026

67378642R00085